RE-USE YOUR CAD: THE

Cover design by Amanda Struz & Rosemary Astheimer
Edited by Sarah Massey-Warren & Michelle Nordwald
Images licensed from 123rf.com and dreamstime.com
Executive Editor: Jennifer Herron



Printed in the United States of America

First Printing: May 2016

ISBN-978-1-532940101

# Re-Use Your CAD

## The ModelCHECK Handbook

Learn how to select, set up, and run ModelCHECK in conjunction with a re-use product definition strategy.

**R. L. ASTHEIMER**
**Action Engineering**

---

*1st Edition – Print Publication*

# Foreword

I've known Rosemary since 2009 when we became colleagues during her days at PTC, and I'm delighted to write a few words of endorsement for this book. Since we all know that forewords are among the most often-skipped sections of a book, you are reading these words either because you are a very detail-oriented and methodical person or because you're still unsure whether reading the rest of this book will be worth your time. No matter which reason it is, you're in luck because your methodical nature will come in handy as you navigate the all-too-numerous configuration files and settings that will confront you during the setup of ModelCHECK and because you hold in your hand a valuable guide that is indeed well worth reading.

As more and more companies are embarking on the journey to becoming a Model-Based Enterprise, the need to validate the quality of their models takes on strategic importance. ModelCHECK can fill a vital role in a broader model validation strategy, which is why many companies are interested in implementing it.

This handbook is an important contribution not because the PTC-provided ModelCHECK Help content is lacking in technical detail on the various configuration files and their functions. That information is indeed adequately covered, which is precisely what any software documentation is intended to do. However, the implementation of any software tool requires more than just a description of its functionality. The technology being implemented needs to fit within a business process and it needs to provide value in support of a business objective.

In this handbook, Rosemary does an excellent job in setting the context for why your company would want to implement ModelCHECK, how it can support your model-based initiatives, and also provides some recommendations on what configuration settings should be used. This handbook follows a logical structure that walks you through everything you need to do step-by-step. There is enough technical detail provided to minimize your need to refer to the official PTC documentation, but it is presented in pleasantly readable prose and with numerous illustrations and diagrams that aid in understanding the concepts. All in all, I'm sure you'll find this to be a valuable resource.

I wish you the best of luck on your MBE journey!

Raphael Nascimento
PTC Product Manager
Host of the "Journey to MBE" podcast available at
www.JourneyToMBE.com

# Prologue

With the arrival of parametric CAD modeling software in the late 1980s, CAD designers were suddenly presented with several different ways to create a solid model that would all result in the same shape. Now that these models had a third dimension to their definition, the approach used during design was different. Rules of thumb were provided by super users of the software, and best modeling practices were defined by the software vendors, but ensuring that these "good" practices were followed was a labor-intensive manual process. This manual process helped drive the need for automated checking of models for "good" modeling practices and compliance with company Standard Operating Procedures (SOP). PTC (Parametric Technology Corporation, as they were called at the time) tackled the problem by acquiring a software program called ModelCHECK that has been used to check models for "bad" modeling practices in an automated fashion for almost two decades now.

At this point, you may wonder: What is a "bad" modeling practice compared to a "good" one? Let's take a minute to look at a specific example of a "bad" modeling practice that ModelCHECK might pick up on, in order to better understand the types of modeling practices that I am talking about. One of my favorite examples of a less than desired method to model a hole is the case of a designer creating an "extrusion" that uses a circular sketch to remove material instead of creating this geometry as a "hole" feature. This classic example makes it easy to see that the resulting geometry looks the same and has the same geometric characteristics as if the hole wizard were used. However, under the hood, the hole feature contains metadata about the hole that the extrusion does not. This is the kind of information that could be re-used downstream by automated manufacturing equipment or automated inspection devices.

As Model-Based Definition (MBD) moves from the early adopter stages into commoditization, authoring a CAD model with geometric and parametric stability, accurate metadata, and re-usable Product and Manufacturing Information (PMI[i]) is imperative. It's essentially the equivalent to the saying "you get out of it what you put into it." More bluntly put, "Garbage in… garbage out."

I'll be the first person to admit that an automated validation software (ModelCHECK, in this case) is not the answer to all of your CAD problems. On the other hand, it can replace some of the tedious menial and manual verifications done by a human today. The first book in the Re-Use Your CAD series, Re-Use Your CAD: The Model-Based CAD Handbook, 1st Edition, lays out a strategy for criteria that your organization may require you to validate.

Another factor that should be considered with ModelCHECK…the price is right. It is "free." Well, nothing in life is ever really free, but ModelCHECK is included when the base Creo Parametric package is purchased, so why not take advantage of it?

After you are kick-started with a training or consultant session, read this handbook to accompany your learning and mastery of ModelCHECK. You'll find that it isn't that difficult to do, and then you will have earned the bragging rights to say that you figured out how to do the configuration yourself. Additionally, this will not be a static system. As your organization matures, so will your need to evolve your checking strategy and how it is accomplished. This is one of those roles that will always need to be filled, so doesn't it seem like a good idea to have someone with internal knowledge of your system (and potentially your intellectual property) that can immediately begin to tackle the necessary changes?

My hope for this book is to de-mystify ModelCHECK for the thousands of Creo (and Pro/ENGINEER) users out there so that model quality overall can continue to be improved, as MBD becomes standard practice.

# Acknowledgements

Thank you to Jennifer who helped me through all of the logistics of publishing my first book. She also made me realize (and reminded me) that it is possible to have a family and a successful career at the same time.

Thank you to Sarah for patiently reading through all of the technical details and reminding me (the gear head who is not a natural born writer!) again and again that "set up" when used as a verb is two words, not one!

Thanks to Michelle for finding the time to help clean up all of the nit-picky things that I became blind to after writing for almost a full year – while she managed four young kids at home! I envy her endless energy!

Thanks to Martin for reviewing the technical details and catching a few minor details that I would have missed.

Thank you to Trevers who put up with me staring at my computer on the couch, in the car, on our flights to and from vacation. Who took "the maniac" grocery shopping or car shopping (insert eye roll here) so that I could write this book and meet my deadline.

And finally to Ayrton, the man who changed my life and my perspective on life in so many ways. It's impossible to understand how much you are loved by someone until you have become a parent yourself. Your best efforts to color in the umbrella on page 14 so that you could be part of my first published book make me so proud. Your unending energy and need to know about **everything** makes me feel young. I love you more!

# Table of Contents

# ABOUT THIS BOOK

I will begin by introducing you to ModelCHECK and all of its files. I will then explain how to edit the files, when they are read and what they do. If you are already familiar with ModelCHECK, have used it in the past or are currently using it, this section will be an easy read. You could probably even skip the first half of this book; however, it never hurts to review and refresh what you think you already know. Who knows, you may come across something new or that you had forgotten about. And as the author I would ask why you bought this book if you aren't willing to take the time to at least skim it?!

The next part of the book will cover how to read the ModelCHECK report and how go about putting together an implementation plan, including how to automate the checking of as many of the practices covered in Re-Use Your CAD: The Model-Based CAD Handbook, 1st Edition as possible. When you read the next section, A Brief ModelCHECK History, you will better understand why it is not possible to automate **every** best practice covered in Re-Use Your CAD: The Model-Based CAD Handbook, 1st Edition. That said, why not automate what you can in order to make your model checking procedure more efficient and less error prone?

The final pages will cover some frequently asked questions and other information that I think you might find helpful as you tackle a ModelCHECK implementation.

Be patient as we cover all of the details first and remember the good 'ol saying:

*"You have to learn to walk before you can run."*

# A Brief ModelCHECK History

ModelCHECK was originally developed by RAND Worldwide[ii], a world leader in providing technology solutions and professional services, founded in 1981 in Ontario, Canada.

By 1984, RAND had partnered with Computervision, Inc.[iii], an early pioneer in Computer Aided Design and Manufacturing (CAD/CAM). Computervision was founded in 1969 by Marty Allen and Philippe Villers, and was headquartered in Bedford, Massachusetts, USA.

It isn't clear exactly when, but sometime in the late 1980s, RAND developed a software application named ModelCHECK to check the quality of models that were created within the Pro/ENGINEER[iv] software. The intent was to verify conformance to design best practices as defined by Parametric Technology Corporation[v] who developed and sold Pro/ENGINEER. By 1988 RAND had partnered with Parametric Technology Corporation to market and support ModelCHECK[vi].

Approximately 10 years later, in 1998, Parametric Technology Corporation acquired Computervision[vii], which was still a partner with RAND at the time.

By 1999, Parametric Technology Corporation had purchased all rights to ModelCHECK[viii], which was being used in the field by approximately 350 customers at that time.

If you are wondering why I keep referring to Parametric Technology Corporation fully spelled out, it wasn't until January of 2013 that Parametric Technology Corportation legally changed its name to PTC Inc.[ix].

Since the acquisition of ModelCHECK and as of the writing of this book, there have not been any drastic changes to the structure of ModelCHECK or how it functions. However, throughout the software versions that have been released over this time period, several new checks have been added to accommodate new functionalities in Pro/ENGINEER and Creo, and modifications have been made to how some checks function (in order to fix a bug, also referred to as a Software Performance Report, or SPR).

So, the good news for you as the reader is that the basic content of this book should remain valid for years to come. The bad news for me as the author is that you will likely only ever need to purchase one copy!

Now that you know that ModelCHECK is almost two decades old, you might be wondering why I wrote this book now. My response to that is answered with a three-letter acronym: MBD. The push towards Model-Based Definition has lots of people concerned about model quality, and for good reason. If the CAD model is going to become the master source authority for manufacturing and inspection information, it should be robust and thorough in its content.

The other reason is because … there's almost a cult following of people who get really excited when they hear the mention of ModelCHECK. As one of the previous Product Managers of the Creo software, I always wanted to give it some love, but there were never enough hours in the day to do so. ModelCHECK seemed to have a permanent place on the back burner of the virtual product development stove. Now that I'm on the other side as a user and have a different perspective, I feel that sharing my knowledge of the product was even more of a necessity.

# PART ONE: A STRATEGIC APPROACH FOR PLANNING A MODELCHECK IMPLEMENTATION

If you can't define it, you can't repeat it…

If you can't repeat it, you can't measure it…

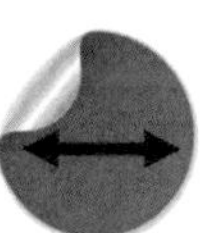

If you can't measure it, you can't improve it…

**And that's bad!!![x]**

# Guided Steps to a Model Checking Implementation Plan

Okay, let's get started! The mouse is in your hand, your finger is hanging over the left mouse button, now what?

Five major groups of tasks must be understood to help make your ModelCHECK implementation a success. They are:

1. Understanding **why** you want to use ModelCHECK
2. **Preparing** ModelCHECK for customization
3. Understanding the different ways to **trigger** ModelCHECK
4. Applying the appropriate settings to **configure** ModelCHECK
5. **Reading** the reports

To get a better appreciation for what we're trying to do here, let me first make you aware that ModelCHECK references a lot of files. It's hard to put an exact number on how many files, because some files contain options that point to other files that could then in turn point to other files and... well, you get the point. The number of files involved in your setup could quickly become very different from another setup based on the specific checks that you've selected to use.

On the next page, take a look at the image that I put together in an attempt to show you the default files included with the standard installation, and keep in mind that this is the starting point. Several other related files and settings exist, but these files are the starting point.

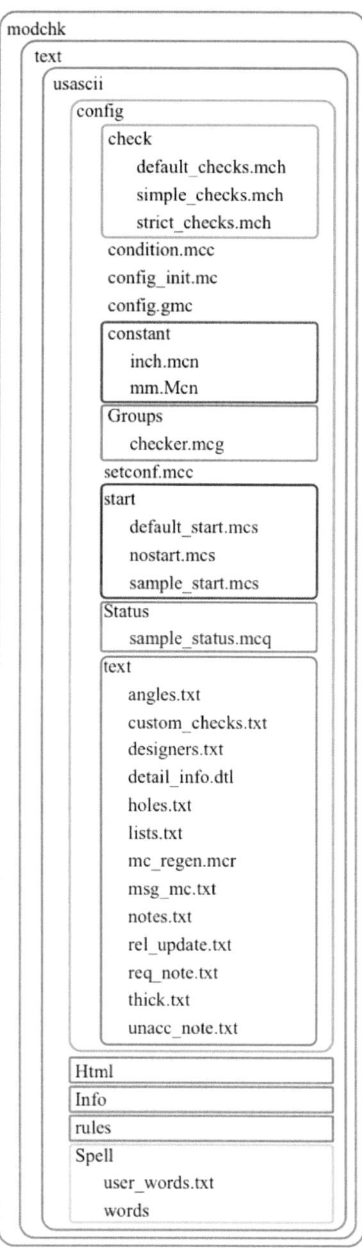


***Default Files Used by a ModelCHECK Installation***

# Understanding Why You Want To Use ModelCHECK

Before we start looking at the actual files involved, we need to understand what we are trying to do so that we can determine who needs to be involved in the setup and ongoing administration and use of ModelCHECK. The bottom line is: get these people involved in the planning stages so they will have the necessary permissions to set up, run and test an installation.

In my years of working with PTC products, I've been asked many times and several different ways how to do X, Y, and Z with Creo (or Pro/E, which is short for Pro/ENGINEER, as we veterans still tend to call it), but it wasn't until I became a parent that I truly began to realize the importance of first asking ourselves the "why" question. Before we jump in and start setting this option and that configuration, we need to stop and ask ourselves, "Why do you want to use ModelCHECK?"

Since the day my toddler could say the word "why," he's been using it over and over (and over!) again, in a non-stop effort to better understand how the world works. I myself have learned a lot from the simple approach that children take, so why shouldn't we do the same in order to validate our intentions, no matter the task? For example, "Why is it raining outside?"

> *Well to be exact, when water becomes warm enough, it evaporates as vapor into the air. When a mass of air quickly cools to its saturation point, the water vapor condenses into clusters of tiny water droplets and frozen water crystals. We call these clusters clouds. Over time, the droplets and crystals that make up a cloud can attract more water to themselves. When water droplets grow heavy enough, gravity pulls them down as raindrops.*[xi]

***Why is it raining outside?***[xii]

That was probably way more information than my toddler was looking for, but technically there is a reason for everything, and since we are working on a technical implementation here, asking why works perfectly.

Of course the answers aren't always so black and white and, depending on who is making the decision, may even involve an element of emotion. "Why is my car red?"

> My car is red because the marketing department of the OEM did some research and determined that 10% of the production of that particular model should be red to facilitate their sales goals. And red happens to be my favorite color. And now that I've owned 4 red cars in a row, the decision of what color to buy next now includes my emotions since having a red car is "my thing!"

Either way, whether the decision was made based on facts or emotions, or a little bit of both, some conclusion had to be reached.

My point in going off on this tangent: when approaching almost anything in life, including considering an implementation of ModelCHECK, doesn't it generally make sense to take a step back and ask why we are considering doing something so that we can have a better understanding of our goal?

*Think like a child—because you'll come up with better ideas and ask better questions.*[xiii]

***WRITE A STATEMENT ABOUT WHY YOU WANT TO USE MODELCHECK***

If you are setting up your own customized implementation, this information can be entered in the Custom ModelCHECK Implementation Checklist found on page 122.

# Preparing ModelCHECK for Customization

## Who Needs to be Involved?

Now that you have decided to take the plunge to configure ModelCHECK, the next thing to consider is whether you have the right people involved. This section will discuss potential roles that you might want to involve, so that they can give their feedback as early in the installation process as possible. Nothing is worse than going through weeks or months of planning only to find out that a key piece of information was left out and you have to return to the drawing board. It is also important to make sure those who need to be involved have the permissions they need to do their part of the job, whether it be running ModelCHECK to generate a report or helping with the configuration.

Each section below is labeled with a job role (or department) typically involved in ModelCHECK administrative tasks or that requires education on how to run ModelCHECK. At a minimum, these next pages will help explain why running ModelCHECK is important and beneficial to the company overall and why it is not a task for a single employee.

## Information Technology (IT)

It is important to know that, by default, ModelCHECK file permissions are locked when Creo is installed. There are two ways to allow users to make changes to the files, which your Information Technology (IT) department may dictate. This section describes the two methods available to set file permissions. Either method will require that changes to the Creo installation are made by someone with administrative privileges.

1. **Operating System** – Permissions can be granted to those who need them through the operating system on which Creo runs. At the time this book was written, Creo was only certified to run on Microsoft Windows. But if you are using this book with an older version of the software, there was once upon a time when Unix platforms were also supported.

2. **ModelCHECK** – One of the ModelCHECK files contains an option that points to a text file containing a list of users authorized to make changes. The details of this option are explained in the *config_init.mc* section of this book. The specific option to be used is called MC_AUTHORIZATION_FILE.

## CAD Designer, Product Designer

Anyone who will be authoring or making changes to a CAD model or assembly needs to understand what will be checked by ModelCHECK and how to properly capture the required elements for the item being checked to be deemed acceptable. Training for these users is best carried out once the installation is set up and running so that they can see the checks working.

## Quality Assurance, Quality Control, Inspection

Quality control needs to understand the design requirements that were specified to be able to release a design as conforming or acceptable quality. This likely means reading at least some of the data output by ModelCHECK to make sure requirements have been met. Training for these users should also be carried out once the installation is up and running.

*"Don't expect what you don't inspect."*[xiv]

## Database Administrator, PDM Administrator

If you have a Product Data Management (PDM) system in use, you will likely want to pass model or assembly level parameters to the PDM system. A PDM administrator needs to be aware of the origin of this metadata to help ensure that the information is passed into the database correctly.

It is quite common that PDM systems are utilized as a gatekeeper to ensure that the quality of the data being entered is "good." This means that the PDM system can be set up to reject the model if a certain set of conditions is not met. For example, if more than two errors were reported by ModelCHECK when it was last run, the file will not be allowed to be checked in. This can be a good thing and a bad thing. On the one hand, implementing such checks in your PDM system tries to ensure that only "good" data is put into the PDM system. On the other hand, if the users who are submitting the data aren't sure of the correct procedures required to make the data good, they may try to find a way around the error, which could then lead to even lower quality data. You can see the virtual dog chasing its tail in this scenario.

## Project Manager, Product Manager

Anyone monitoring a project or product may also benefit from knowing how to run ModelCHECK on a CAD file. Alternatively, having access to the results of the last ModelCHECK run or a summary of the reports may also be sufficient.

***IDENTIFY THE APPROPRIATE TEAM MEMBERS NECESSARY TO PUT AN IMPLEMENTATION IN PLACE, RUN AND ADMINISTER MODELCHECK.***

If you are setting up your own customized implementation, this information can be entered in the ModelCHECK Implementation Checklist beginning on page 123.

## What Do You Want To Check For?

The hardest part of setting up a model checking software is not how you will do it, but deciding what you want to check in the first place. Many organizations believe they have a thorough list, but when the fingers hit the keyboard, those tasked with automating the checking pause with their index finger over the mouse wondering, well... "What feature should I begin checking? Why should I be checking it? How do I check it?"

Depending on how well you understand your organization or group, you may already have a general idea of the types of things that end up causing rework or another iteration of the design process. If you are new to your organization or don't have a gut feeling of the types of things that occur on a regular basis, another option is to sift through a history of records or reviews of previous projects or programs, looking for trends. If you already have a PDM system in place, this history may already be stored in an organized format that makes it even easier to sort through. The point is, if you can identify why parts fail inspection or require rework after being produced, then you have a good starting point for understanding what your goals might be, which will then determine how to measure with ModelCHECK.

*"Avoiding high cost of engineering changes is a prime driver for going to MBD and model checking. A well-thought-out product design verification strategy that combines automated and human checks will reduce rework and scrap costs." – Jennifer Herron*

Let's take a closer look at the basic definition of ModelCHECK, which will help us lay out a structure that we can follow to begin setting up a plan. This is by no means an all-inclusive list, so don't feel that you are cornered into only checking for the types of things listed here. Think outside of the box. Poll your team and discover which checks are required for all files, and what might vary for different files or products.

**ModelCHECK is:**

A configurable **set of checks** used to analyze:

- Parts
- Assemblies
- Drawings

In order **to**:

- Verify standards compliance
- Check for "good" modeling practice
- Ensure a complete data set

Which will **promote** the effectiveness of downstream data by:

- Checking critical features and/or areas of a part, assembly or drawing

For **no additional charge** since it is integrated into base Creo licensing!

Based on the above definition you can begin to think about a few things and maybe even answer a few questions, such as:

- What types of files do I want to check?
    - Parts, .prt files
    - Assemblies, .asm files
    - Drawings, .dwg files - If you are fully embracing MBD, you might not even create drawing files anymore, which takes them out of the list of things to be checked.

- What are you checking for?
    - Conformance to a standard?
    - Whether designers are turning out good quality models?
    - Missing or incomplete data within a file?

- Who will be doing the checking?
    - The person creating the CAD model?
    - Someone tasked to check the model before it is released?

Identifying the above information might help you figure out your goals. Remember, setting up ModelCHECK as a quality checking mechanism will only be as good as how thoughtful you were in deciding what to check. The list below includes a few examples of goals that you might come up with.

**Example goals that you might set:**

- Ensure that parameters exist in a model or assembly:
    - Ensure that parameter values are not empty.
    - Ensure that parameter values follow the correct syntax.

- Check for basic poor modeling practices such as:
    - Adding a finishing feature such as a round, chamfer or draft feature to a model early in the design process (they typically should be added at the end of the design).
    - Creating a hole as an extrusion instead of a “hole” feature.

- Check notes for spelling mistakes.

- Ensure that notes are only those defined in a library of acceptable notes set up by your company.

***SET ONE OR MORE MEASUREABLE GOALS TO BE ACHIEVED BY YOUR MODELCHECK IMPLEMENTATION***

The above action item will make it clear why you are going through the effort of setting up a ModelCHECK installation and will ensure that you have a way to quantitatively measure your return on investment. It will also help you identify available checks that you may want to use as we review them in detail in the next section.

*“If you can’t measure it, you can’t improve it.”* [xv]

To many people, ModelCHECK seems to offer the illusion of a solution to all your modeling problems. A panacea of hope, but that isn’t realistic. Can it help improve your model quality? Absolutely. But keep in mind, just as getting a good quality CAD model requires following best modeling practices, so does getting usable data from a carefully thought out ModelCHECK implementation.

Once your “checking” checklist is generated, then it is fairly straightforward to quantify the return on investment that ModelCHECK can bring by identifying those checklist items that can be automated. Keep in mind, most conditions you set in ModelCHECK must be either pass or fail, and subsequently, the automated checklist criteria must also be set to return a pass or fail answer. We know the user interface leaves much to be desired, but if you are able to manipulate the files with my guidance, the benefit of automating a portion of the checker’s workload is significant.

# PART TWO: UNDERSTANDING HOW TO TRIGGER MODELCHECK

2

It might seem like this section is out of order, but in order to configure the checks, you have to first make sure that ModelCHECK is turned on and understand how it is run.

## Turn ModelCHECK "On"

If you've been using Creo for any amount of time, you probably already know that, at startup, Creo reads a system configuration file named *config.pro* to set up the configuration of Creo when it launches. In the case where more than one *config.pro* file is found (in different directories) the last option that was read will be the value used. It's also possible that a *config.sup* file has been put in place by an administrator, which could cause the value of this option to be overridden.

The option that we are concerned with in the *config.pro* file is named '*modelcheck_enabled'* and it must be set to '*yes'* (the default value), in order to use ModelCHECK.

***ENSURE THAT THE MODELCHECK_ENABLED OPTION IN THE CONFIG.SUP OR CONFIG.PRO FILE REMAINS SET TO 'YES'.***

Creo will look for the *config.pro* file in three locations, in the following order:

1. In the Creo installation directory, under the "Common Files\text" directory:

   The default will look something like this:

   *C:\Program Files\PTC\Creo 3.0\M040\Common Files\text\config.pro*

2. In the directory to which the 'HOME' environment variable is pointing:

   In the Control Panel, under Administrative Settings, you will be able to see a list of variables that have been set. The procedure to get to this information is a bit different depending on your Windows OS version. If you don't know how to locate this, your IT department can help, or you can try to find it using one of the steps listed on the next page.

3. In the directory in which the Creo shortcut was set to start:

   If a shortcut exists on the desktop to launch Creo, it can hold a 'start in' directory. Following the steps in the box below to locate this information.

**Finding the 'Start In' Directory for a Shortcut on Windows 7**

- Right mouse click on the shortcut that you use to startup Creo and choose **Properties**
- Look at the path found in the "Start In" field

**Finding the 'Start In' Directory for a Shortcut on Windows 8**

- Right mouse click on the shortcut that you use to start up Creo and choose **Open file location**. A windows explorer will be opened in the directory where Creo is set to look when it starts.

This page gives detailed pick-and-click steps so that you can change the HOME environment variable discussed in the previous paragraph.

**How to find the value of the 'HOME' Environment Variable**

*Note that you may not be able to make any changes to the variable if you don't have write permissions on the file, but you should be able to at least see where it is pointing.*

**Finding the 'HOME' Environment Variable on Windows 7**[xvi]

- Click the Start button, right click on Computer, click on Properties
- Click on the Advanced tab
- Click the Environment Variables button in the bottom right corner
- Locate the name of the user variable or system variable named 'HOME' and change it

The alternative (which is debatably easier to get to the result, yet more difficult to read) is to:

- Open up a command line
- Type "echo %HOME%"
- If the variable is set, the paths that are associated to it will be listed

**Finding the 'HOME' Environment Variable on Windows 8**[xvii]

- In Windows Search, select Settings (Windows Key + W) and search for "Environment Variables"
- Choose "Edit environment variable for your account" or "Edit the system environment variables," based on where you want to apply the change
- Locate the name of the user variable or system variable named HOME and change it

The alternative (which is debatably easier to get to the result, yet more difficult to read) is to:

- Open up a command line
- Type "echo %HOME%"
- If the variable is set, the paths that are associated to it will be listed

With the *config.pro* option MODELCHECK_ENABLED set to *'yes'*, another *config.pro* option named MODELCHECK_DIR can be utilized. If this option has been added, its value will be set to a file path. This allows the ModelCHECK files to be placed in an alternate directory other than the default location (*C:\Program Files\PTC\Creo 3.0\M040\Common Files\modchk\).* If this option has not been set, Creo will look for the files in the default location.

As an example of 'HOME' file location, you might set the MODELCHECK_DIR option to something like this:

*C:\Program Files\PTC\Creo 3.0\M040\ModelCHECK_MBD_Setup*

# Running ModelCHECK

Okay, you are probably thinking, "enough planning already." Let's get down to the details of how it is done! Before I unveil all of the files that hold the magic switches to determine what ModelCHECK will do, you must first be familiar with the four ways that ModelCHECK can be run. The reason you must be familiar with the four different ModelCHECK execution methods is because many of the checks are not turned on. Rather, ModelCHECK is not configured as you might expect it to be out of the box.

Bear with me for a few more paragraphs while I first explain the four modes. Then when we get to the details of the files, you will be able to recognize these four modes, and all "how to run" ModelCHECK will make more sense.

## Interactive Mode

This method is called interactive mode because it relies on the user to manually run ModelCHECK. In Creo, from the *File* menu, select *Prepare*, and then *ModelCHECK Interactive*. This method is typically used during the design process. Designers can run ModelCHECK on their work as they go to catch potential problems before they get too far down the path towards a final product. Because Interactive Mode engages and runs ModelCHECK immediately, this mode is also an effective method for performing a quick spot-check review of another user's model before working on it.

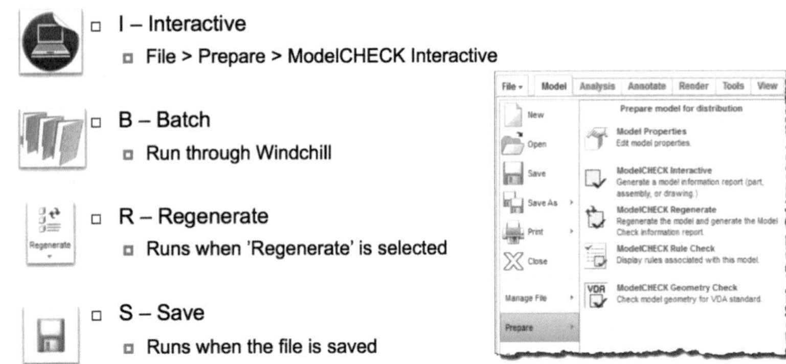


***Overview of the ModelCHECK Modes and Image of the File Prepare Menu in Creo***

## Batch Mode

This method runs ModelCHECK when a batch process (usually run outside of Creo with no User Interface) is run. Often when a large number of files need to be processed in some way, an automated process called a "batch" is set up to run. The process will perform its task on each file, one after another until all of the files have been processed. Processes can be set up to do just about anything, but common scenarios include:

- Opening the file and performing a "Save As" in order to save files out to another format

- Running scripts that automate a task that is tedious, and probably highly error prone, when done manually.

To make the most efficient use of this processing time, you may want to also run ModelCHECK as part of the batch process. Let me explain why this can be beneficial by taking a look at a theoretical example.

Let's say the average time to load a model or assembly in Creo is 30 seconds. You then want to run a script on the file, which will take 10 seconds and finally translate the file to .STP format, which will take an additional 20 seconds. The total time needed to perform these three steps (open the model, run the script and translate the file) would be 60 seconds.

Now let's say that we open the file at a later time just to run ModelCHECK, which let's say also takes 30 seconds. The total time for this task is also 60 seconds (remember we have to again wait for the file to open for 30 seconds).

Compare the time required to run each of these scenarios. Opening the file just to run ModelCHECK adds an additional 30 seconds to the process, so if you can integrate ModelCHECK into the batch process, you will eliminate opening the file a second time at a later date at the cost of another 30 seconds.

If you are a visual learner, take a look at the image on the next page to see that running the batch process and the ModelCHECK process separately will take a total of 120 seconds, but by running ModelCHECK as a supplemental check to the batch process you will save 30 seconds. This means you will be able to process four times the models compared to the total time needed if you were to go back and run ModelCHECK separately at a later time.

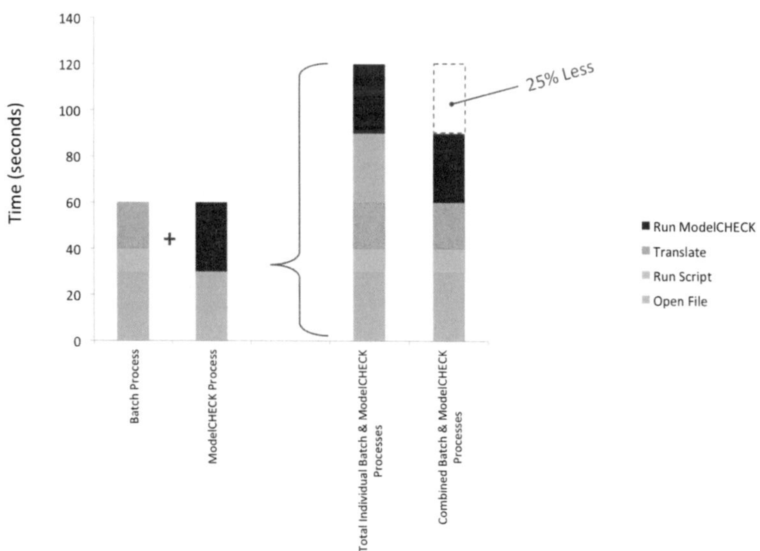


***Running ModelCHECK in Conjunction with a Batch Process is More Efficient than Running the Two Tasks Separately***

Hopefully this makes some sense and you understand my point that you may as well get as much bang for your buck when you can. Who doesn't love a bargain!

## Regenerate Mode

This method runs ModelCHECK whenever a model is regenerated. While PTC recommends this mode as the most effective way to run ModelCHECK, I believe that it can slow down the design process considerably because it runs ModelCHECK every time something in the model is modified and requires regeneration. If you've used Creo for any significant amount of time, you know this can happen quite often and, if your model is complex, can leave you sitting staring at the hourglass for a while.

If you have a PDM system in place, and it is set up to check the ModelCHECK results before allowing a file to be checked in, I wouldn't recommend spending the time waiting for ModelCHECK to run with each regeneration of the model.

## Save Mode

This method runs ModelCHECK whenever a model is saved. This might be the sweet spot if you are concerned about the productivity hit that running ModelCHECK on regeneration might have, but you aren't 100% confident that you want to rely on your designer to remember to execute the checking process.

As I mentioned above, of the four methods available, I always prefer running in interactive mode because I think it's best to get feedback on the model right away, while the portion of the design that you are working on and its design intent are fresh in the designer's mind. However, I also recognize that no one solution works for everyone. You might even want to set different levels of checking based on which mode ModelCHECK runs in. For example, Save Mode might run very basic checks, but Interactive Mode could run more detailed checks since you have immediate access to the details of the check's results.

# Setting the ModelCHECK Options

The point of introducing the four modes above is that when you see the options listed in many of the files, they will be followed by a sequence of (usually four) letters. The possible values of those four letters represent the setting for the particular option being set in each of the four modes. The letters are usually Y, N, A, E or W. Let me explain what these letters mean and walk you through an example.

## Y (Yes)

As you might have guessed, Y is for YES, and it means that you want to run that check.

## N (No)

And N is for NO and means that you don't want to run that check.

## A (Ask)

The less obvious option A means ASK the user when appropriate if they would like to run that check.

## E (Error)

This option won't be used until we get further down into the files, (page 68 to be exact) but it means that if the condition specified is met, the ModelCHECK report will show an ERROR.

## W (Warning)

This option also won't be used right away (also on page 68), but it means that if the condition specified is met, the ModelCHECK report will show a WARNING.

If you are scratching your head trying to understand what I was trying to describe above, take a look at the figure below. It shows the contents of one of the files used by ModelCHECK that will use the "four letter" setting that I was just mentioning. Keep in mind that any line beginning with ! or # is considered a comment and will not be read by ModelCHECK.

Take a look at the text inside the red box in the image below that starts with MODE_RUN. This is an example of an option in this file to be read. It is first followed by two letters (YN) that describe the possible values that can be set for the four modes (Interactive, Batch, Regenerate and Save), which are then next in line.

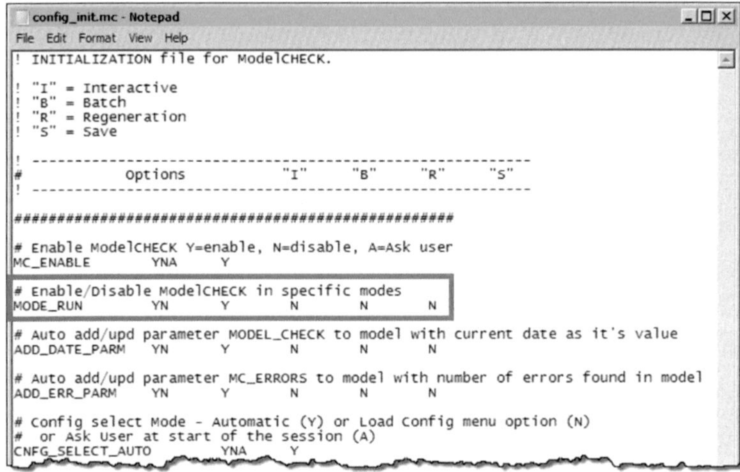


***Example ModelCHECK File in Notepad***

Taking a closer look at the contents of this file, let's write out in plain English what the file will do:

- Enable ModelCHECK to be run
- Set the option MODE_RUN to YES for Interactive mode and to NO for Batch, Regenerate and Save modes
- Set the option ADD_DATE_PARM to YES for Interactive mode and to NO for Batch, Regenerate and Save modes
- Set the option ADD_ERR_PARM to YES for Interactive and Regenerate modes and to NO for Batch and Save modes
- Set the option CNFG_SELECT_AUTO to YES

The last option set in this file is a bit different. It only has one value. How can you be sure that there is only supposed to be one value? You can take a look at the default files to see how they are set up. If there is only one value, then that is probably what ModelCHECK is expecting; however, the possibility always exists that someone edited the file and accidentally removed the other three values. You can refer to the documentation of that option if you want to double check. If you don't see a table in the documentation describing the suggested settings for the four modes, then indeed only one option needs to be set.

***TO ENSURE CONSISTENCY WITHIN YOUR ORGANIZATION, COORDINATE WITH I.T. ADMINISTRATION TO SYNCHRONIZE FILE CONTENT ACROSS ALL CREO CLIENT MACHINES.***

# 3 PART THREE: CONFIGURING MODELCHECK

You've now identified why you are going down the path towards a ModelCHECK implementation, and you have identified whom you will need on board to set up the installation. The next logical step is to get these people to help you determine **what** to check based on the defined goals. This means getting the team assembled as early in the process as possible so that everyone can become familiar with the files and process involved in setting up ModelCHECK.

What constitutes a "good" model? No one answer applies to this question because it is different for every industry, even different for companies within the same industry and possibly even different for departments within the same organization. What do I mean by that? Consider a medical devices company manufacturing a stent to be implanted inside the aorta of a human heart (to prevent an aneurysm from bursting). This device will have to meet far different criteria than components created by the same company to be used in a reconstructive system implanted in the spine to provide support for a spinal fusion.

Of critical importance is the definition of the checks, and that the authors of models understand each criteria and how to comply with the checks running in ModelCHECK. Otherwise, you and your designers will reach a stalemate.

***AS YOU PROCEED THROUGH THIS SECTION OF THE BOOK, REFER BACK TO THE DEFINITION OF MODELCHECK (PAGE 19) AND THE LIST OF THINGS THAT YOU'D LIKE TO CHECK (PAGE 123). THIS WILL HELP YOU SELECT THE MOST BENEFICIAL CHECKS FOR YOUR IMPLEMENTATION.***

# The ModelCHECK Files

Now that we have a better understanding of what we are trying to do, who needs to be involved and the basic syntax of the files, it should be restated: a **lot** of files are involved with the setup of ModelCHECK.

We'll take a look at the files one by one and, by the end of this section, you will have a better understanding of how things work and will be able to imagine how complex a ModelCHECK installation can become if you begin adding lots of intelligent checks.

Because I am a visual learner, along with 65 percent of the population[xviii], when I began to try to unravel the fur ball of available options, I wanted a visual representation of the files and their hierarchy. As it turns out, this was just the outer layer of the virtual fur ball! Many more connections exist between files and between options within files, but let's not worry about that just yet. We'll get to that when the time comes.

The figure on the next page shows a visual representation of how the files are read. The Creo Configuration file (*config.pro* or *config.sup*, as was discussed on page 22) is the first file involved with configuring ModelCHECK. The remaining parts of the figure refer to files that are specific to ModelCHECK; the next section will cover the contents of these files in detail.

Remember to refer to the section of this book titled "Who Needs to be Involved?" and specifically to the paragraph discussing the Information Technology role in order to make sure users making modifications have the permissions they need. **Administrators of ModelCHECK must read through this section!**

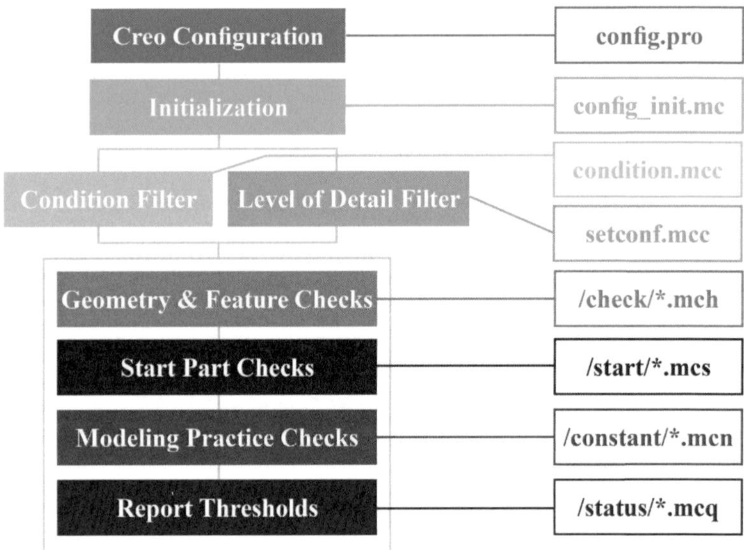


***Visual Representation of ModelCHECK Files***

The colors used in the chart above will be re-used throughout the book to help you identify what file is being discussed and as a cue to help you understand where you are in the process.

# Modifying the ModelCHECK Files

Before we take a look at the details of the files (we're almost there – I promise!), I want you to be aware of the two different methods available to make changes to the files.

## 1. User Interface

Under File > Options > Environment is an option to launch a User Interface that will allow you to make changes to the ModelCHECK files, as shown below.

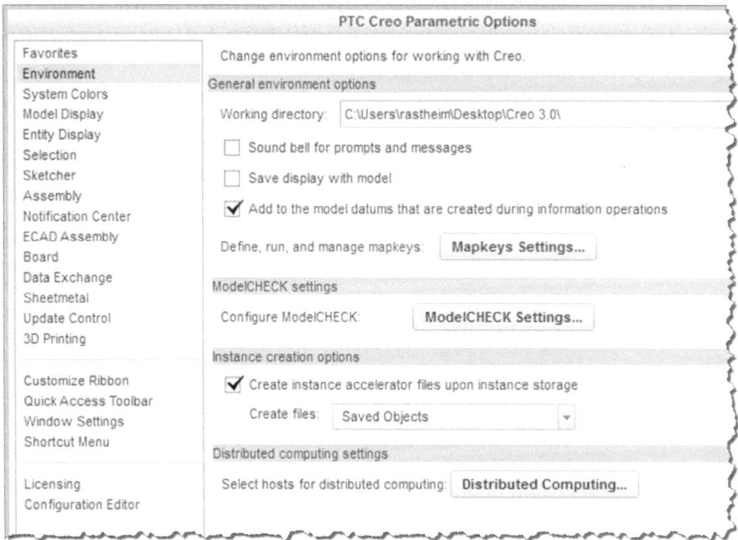


***Launching the User Interface for ModelCHECK in Creo 3.0***

While this method lays out the available options in an organized and easy-to-read manner, it does have a caveat… you cannot perform a search for a function. Not being able to perform a search can make it tedious to locate the option you want in a long list of available checks.

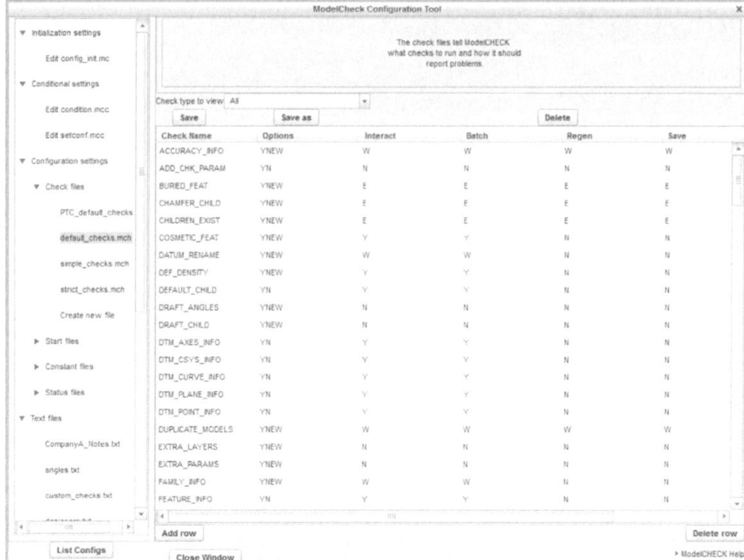


***ModelCHECK User Interface in Creo 3.0***

Since I generally have an idea of the option that I'm looking for, or at least some of the keywords that will have been used to name the checks, I like to search, which leads me to my favorite alternative method, using a text editor.

## 2. Text Editor

All of the files that ModelCHECK reads are simple text files that can be edited with the most basic text editors, such as Notepad or Wordpad. The downside to the text editor approach is that it is not always the easiest format to read because the columns of information often aren't well aligned.

The following image shows one not-so-bad example of how the available options and set options don't align perfectly because of the simple fact that the names of the options shown here all have a different number of characters in each of their names. If this really keeps you up at night, you could add white space in between all options so that everything lines up for easier reading, but this is not necessary in order for the file to be read by the machine. However, I get the need to have visual order as you slog through the files.

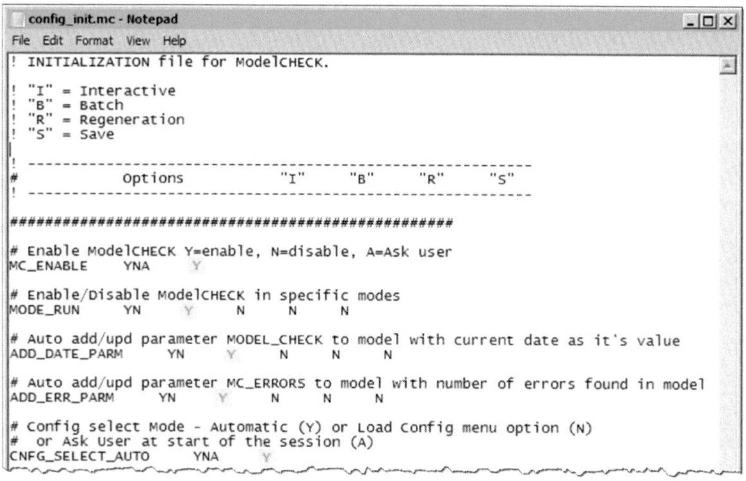

```
config_init.mc - Notepad
File  Edit  Format  View  Help
! INITIALIZATION file for ModelCHECK.

! "I" = Interactive
! "B" = Batch
! "R" = Regeneration
! "S" = Save

! ---------------------------------------------------------
#           Options            "I"      "B"      "R"      "S"
! ---------------------------------------------------------

####################################################

# Enable ModelCHECK Y=enable, N=disable, A=Ask user
MC_ENABLE     YNA      Y

# Enable/Disable ModelCHECK in specific modes
MODE_RUN      YN      Y      N      N      N

# Auto add/upd parameter MODEL_CHECK to model with current date as it's value
ADD_DATE_PARM      YN      Y      N      N      N

# Auto add/upd parameter MC_ERRORS to model with number of errors found in model
ADD_ERR_PARM      YN      Y      N      N      N

# Config select Mode - Automatic (Y) or Load Config menu option (N)
#  or Ask User at start of the session (A)
CNFG_SELECT_AUTO      YNA      Y
```

***A Text Editor can be Used to Edit and Search ModelCHECK Files, Although the Arrangement of the Options is Not as Organized and Easy to Read as the User Interface in Creo***

# CONFIG_INIT.MC

Congratulations! You finally made it to the first ModelCHECK file. It is named config_init.mc and is what I refer to as the initialization file. It is shown in the figure in the light green rectangle.

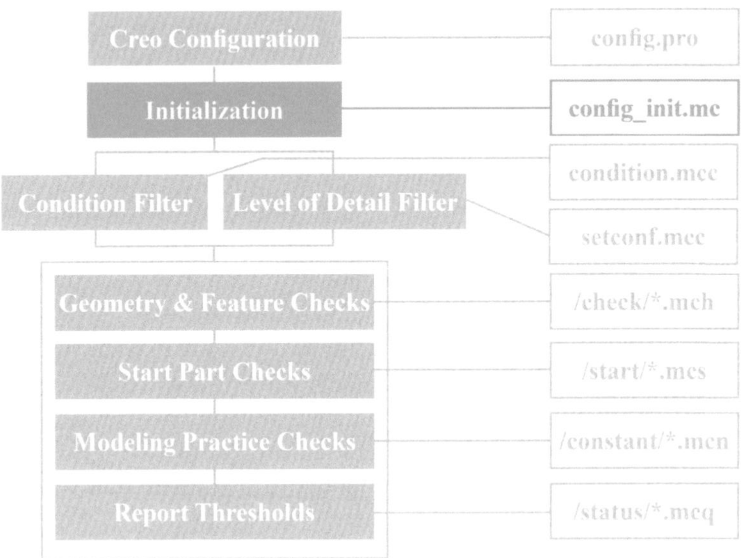


***Focusing on the Initialization File within the Visual Representation of ModelCHECK Files***

This file is read when Creo starts, and receives the message that ModelCHECK has been turned on by the option in the *config.pro* file. Based on its name, you may have already guessed that this is the very first ModelCHECK-specific file read in order to initialize the ModelCHECK setup.

The first thing to be aware of is that, because this is an initialization file, it is only read at startup.

***WHEN CHANGES ARE MADE TO THE CONFIG_INIT.MC FILE, RESTART CREO IN ORDER FOR THEM TO TAKE EFFECT.***

The initialization file contains about 40 options (but several options are related to Unix installations of Creo or Pro/INTRALINK, which are no longer valid, so depending on what you count as an applicable or valid option, you may come up with a different number).

Let's take a look at an actual file to make sure it is clear what the combinations of options are telling ModelCHECK. Remember: any lines starting with ! or # are for comment purposes and are not read by the system.

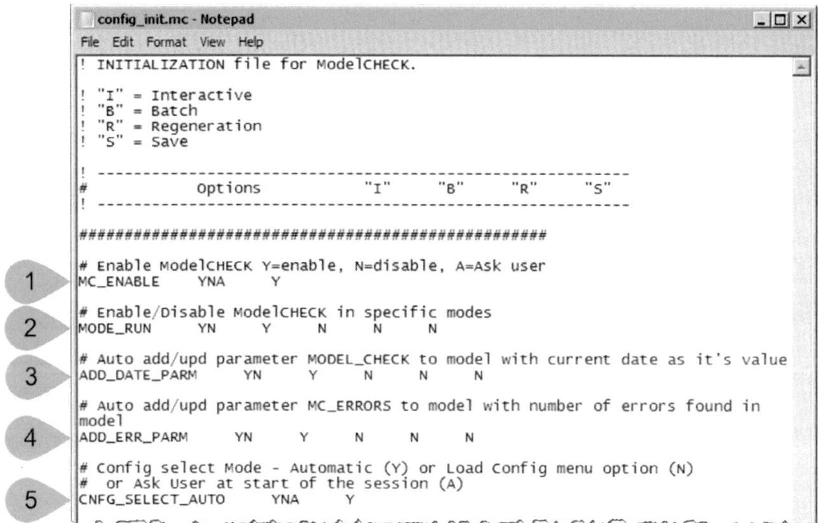


```
! INITIALIZATION file for ModelCHECK.

! "I" = Interactive
! "B" = Batch
! "R" = Regeneration
! "S" = Save

! ----------------------------------------------------------
#            Options            "I"      "B"      "R"      "S"
! ----------------------------------------------------------

####################################################

# Enable ModelCHECK Y=enable, N=disable, A=Ask user
MC_ENABLE      YNA      Y

# Enable/Disable ModelCHECK in specific modes
MODE_RUN       YN       Y       N       N       N

# Auto add/upd parameter MODEL_CHECK to model with current date as it's value
ADD_DATE_PARM       YN       Y       N       N       N

# Auto add/upd parameter MC_ERRORS to model with number of errors found in
model
ADD_ERR_PARM       YN       Y       N       N       N

# Config select Mode - Automatic (Y) or Load Config menu option (N)
#  or Ask User at start of the session (A)
CNFG_SELECT_AUTO       YNA       Y
```

***The ModelCHECK config_init.mc File in Notepad***

The first line read in the file starts with **MC_ENABLE**, and it is what turns ModelCHECK on.

The second line read starts with **MODE_RUN**, and it tells ModelCHECK to allow ModelCHECK to run in Interactive and Batch mode, but not in Regenerate or Save modes.

The third line read starts with **ADD_DATE_PARM**, and it tells ModelCHECK to add or update a date parameter when it is run in Interactive or Regenerate mode, but not in Batch or Save modes.

The fourth line read starts with **ADD_ERR_PARM**, and it tells ModelCHECK to add or update a parameter set to the number of errors found when it is run in Interactive mode, but not in Batch, Regenerate or Save modes.

The fifth line starts with **CNFG_SELECT_AUTO**, and it tells ModelCHECK to ask the user which configuration file should be read.

The remainder of this section covers the options that are relevant to check MBD data sets. They are listed in the order that you will find them in the file, followed by an explanation of what I think they should be set to and why. A few options contain only one setting and, when this is applicable, the default value will show only one letter. I have inserted additional comments where needed.

Here is an example of the layout for the remainder of this section. The 'OPTION NAME' is the variable in the ModelCHECK files. Available Settings are identified with '<VALUES>' in the middle column and 'SETTING FOR "I B R S" illustrates an example of recommended values for each respective mode: Interactive, Batch, Regenerative or Save. Followed by this setup guide is further discussion on the What, Why, and Comments on the significance of, and recommended settings for, each OPTION NAME. Note that I do not make a comment on all options. Phew, aren't you relieved?

Good luck!

Below is an example of the syntax that will be used to guide you through each setting throughout this book. The content may vary between files but the layout will remain similar.

| **OPTION** | **<VALUES>** | **SETTINGS FOR "I B R S"** |
|---|---|---|
| What: | Description of what the option does. | |
| Why: | Why you might want to use or not use this option. | |
| Comments: | Further comments. | |

***Example of the Layout Used in to Describe Each Setting in the File Explained in This Section***

## Settings and Recommendations

### MC_ENABLE <YNA> Y

What: This determines if ModelCHECK will be allowed to run or not.

Why: While you do want to allow some level of checking of your CAD data, there may be some instances where you want to limit or disable users from running ModelCHECK. In general, you will want to either set this to Y to enable it, or A to ask the user; otherwise this book won't do you much good!

Comments: Since this is the master switch to turn ModelCHECK on or off, it isn't relevant to give it a value for each of the four conditions; hence only one value is needed.

### MODE_RUN <YNA> Y N N N

What: This determines the operating modes where ModelCHECK will be allowed to run.

Why: This is another mechanism for use to limit or disable users from running ModelCHECK in specific modes.

### MC_AUTHORIZATION_FILE <YN> Y

What: If this option is set to Y, an external file named external_access.txt will be used to control who has access to modify the ModelCHECK configuration files.

In the section titled "Who Needs to be Involved?", we discussed the two different ways to give permissions to the appropriate users to make modifications to ModelCHECK.

In order for the external file to be read, an environment variable named MC_DIR must be set. The file is then expected to be in a folder named "text" that is inside the directory to which the MC_DIR environment variable has been set.

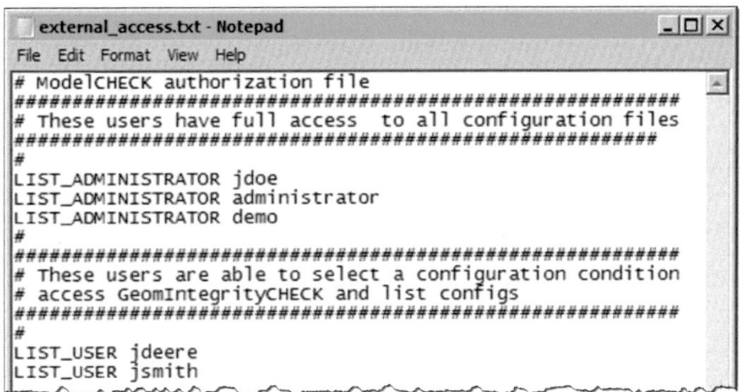

```
# ModelCHECK authorization file
##################################################################
# These users have full access  to all configuration files
##################################################################
#
LIST_ADMINISTRATOR jdoe
LIST_ADMINISTRATOR administrator
LIST_ADMINISTRATOR demo
#
##################################################################
# These users are able to select a configuration condition
# access GeomIntegrityCHECK and list configs
##################################################################
#
LIST_USER jdeere
LIST_USER jsmith
```

***The external_access.txt File that Sets User Permissions for Modifying ModelCHECK files***

The external_access.txt file will follow the format shown in the previous image and will contain a list of users who are allowed to make modifications to the files that determine how ModelCHECK is configured (and who is not allowed to make modifications). A username following the option LIST_ADMINISTRATOR will have full access to all configuration files, while usernames that follow the option LIST_USER will be able to choose a configuration condition, access GeomIntegrityCHECK, and utilize the configuration options that have been set.

Why: You may find it easier to manage who has permissions to change ModelCHECK through this simple text file, rather than setting permissions at the operating system level.

## ADD_DATE_PARM <YN> Y N N N

What: This option adds a string parameter (if it doesn't already exist) or updates the existing string parameter named MODEL_CHECK to the current date when ModelCHECK is run.

Why: It's good to know the last time a check was run.

Comments: The value of the parameter will be in the following format:
*Sun Jan 01 2001 12:34:56 GMT-0500*

## ADD_ERR_PARM <YN> Y N N N

What: This option adds an integer parameter (if it doesn't already exist) or updates the existing integer parameter named MC_ERRORS that is set to the number of errors that were found in the model when ModelCHECK was last run.

Why: This is analogous to a test score of the model and is very often used as a gatekeeper for entry into the PLM in order to keep models not up to standard out of the database. Every time ModelCHECK is run, this parameter will update. Don't worry – even if you find some errors, you will still have a chance to fix everything so all of your hard work can be checked in!

## ADD_CONFIG_PARM <YN> Y N N N

What: This option adds a string parameter (if it doesn't already exist) or updates the existing string parameter named MC_CONFIG that is set to the name of the configuration file used when ModelCHECK was last run.

Why: It's never a bad idea to keep such information around. Perhaps it will explain a trend in failed check-ins if there was a standards change at one point in time, or might help explain why a file wasn't checked into the PDM system as expected.

Comments: The value of the parameter will be in the following format:
*check/default_checks.mch, start/nostart.mcs, constant/inch.mcn*

## DRW_SHEET_ALL <YN> Y N N N

What: This option determines if all of the sheets in a drawing are checked.

Why: I hesitated about putting this option in here at all because I am hopeful that Model-Based Definition will someday eliminate the need to check drawings, but I realize that drawings still do exist and will likely persist for a while.

## CNFG_SELECT_AUTO <YNA> Y N N A

What: This will determine if the conditions set in the *condition.mcc* file will be read, or whether you can choose to load the options in the Load Config drop-down list in the user interface. Refer to the section that covers the *setconf.mcc* file starting on page 61 for the detailed information on the contents of this file.

Why: What you set this option to basically depends on how much you trust the people who will be running ModelCHECK. If this option is set to Y, the configuration files will automatically be selected based on a set of conditions that have been specified. When set to N, the user will have to select the configuration to be used from a drop-down list in the User Interface.

## MC_VDA_RUN <YN> Y

What: VDA refers to Verband der Automobilindustrie 4955 design specification, which is a standard for quality in the automotive industry.

When the MC_VDA_RUN option is set to Y, a tool called the GeomIntegrityCHECK is run, which ensures that the VDA standards are met and that the model does not contain any geometry that could potentially make it difficult to work with downstream.

Why: There are a lot of things looked at in an attempt to check for good basic modeling practices with the setting of this option. As you might expect by now, I say why not take advantage of it!

Comments: When running GeomIntegrityCheck directly (using File > Prepare > ModelCHECK Geometry Check) a file named *config.gmc* determines the thresholds set for each of the checks. However, in order for VDA results to be read during ModelCHECK interactive mode and displayed in the report, the options have to be added to the **.mch* file.

***OPTIONS THAT SET THE THRESHOLDS FOR THE VDA CHECKS HAVE TO BE ADDED TO THE *.MCH FILE THAT IS BEING READ.***

## MU_ENABLED <YN> Y

What: This allows you to update errors that are found by ModelCHECK without any manual interaction. This is done through options that are set in the .mcs file, which we will cover on page 79.

Why: Why not automatically update information if it can be done?! Manual updates take more time and are prone to human error.

## SHOW_REPORT <YN> Y N N N

What: This will determine if the generated report will show in the embedded Creo browser.

Why: Unless you have another mechanism in place to record and display the results of the check, this is the simplest option.

Before moving on, let's take a look at a visual representation of the ModelCHECK installation files that we've discussed to this point, focusing on the config_init.mc file and the options that require it to point to other files.

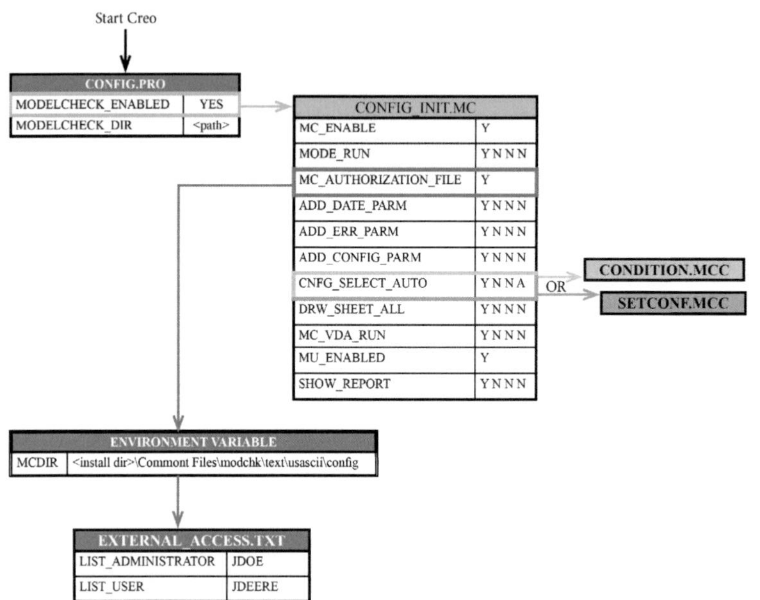


***Mapping of ModelCHECK Files to config_init.mc***

# CONDITION.MCC

The file named *condition.mcc* is what I refer to as the Condition Filter, and it is the next file that is read if the option *CNFG_SELECT_AUTO* is set to *Y* in the config_init.mc file.

Note, that if *CNFG_SELECT_AUTO* is set to *N* or *A* in the config_init.mc file, then the *senconf.mcc* file, which is discussed in the next section, will be read instead.

The condition filter will define the conditions or circumstances that need to occur, using equality statements, in order for a specified set of files to be read.

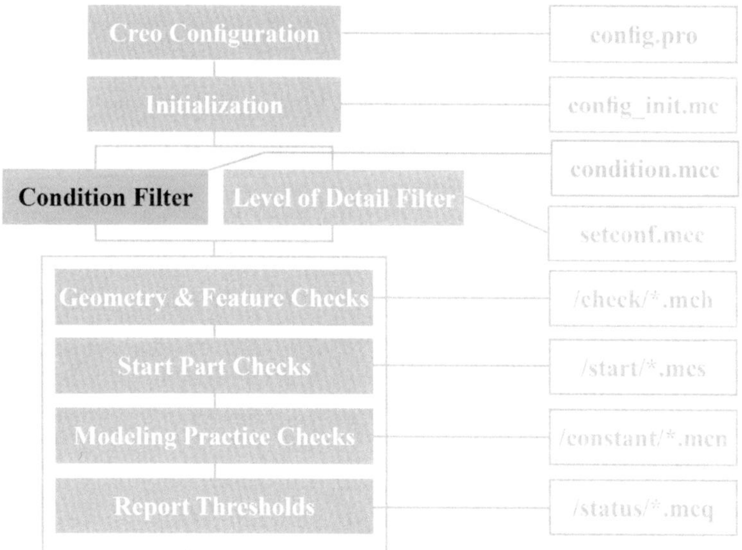


***Visual Representation of ModelCHECK File Focusing on the Condition Filter File***

It is impossible to recommend a default set of values that will work for all organizations in all circumstances, so instead I'll discuss the basic syntax of how the conditions are set up, as well as the operators and parameters that are available to be used, and I will give a few examples of how they might be used.

The basic syntax of the lines in the file is as follows:

> *<Conditional Statement>* config = *<list of files>*
>
> where:
>
> *<Conditional Statement>* is one or more checks against the value of a model parameter or system parameter
>
> and:
>
> *<list of files>* is the list of ModelCHECK configuration files that will be read if the conditional statement is found to be true.

There can be one or more conditional statements in the file, but there is always only one *ELSE* statement at the end. If no *IF* statement is found to be true, then the *ELSE* statement will be read.

Each conditional statement should be in one line. The image below shows the statements wrapping to the next line so that the image could fit on the page, but the statements must not contain carriage returns. If you do have carriage returns, you will not get any errors and the condition will just not work!

Scratching your head yet? This might not make much sense if you haven't seen the contents of the file before, so let's take a look at the actual file. The following is an example of a simple *condition.mcc* file:

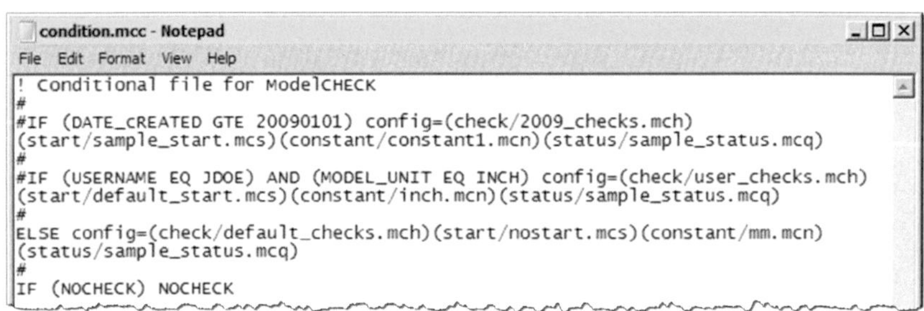


```
! Conditional file for ModelCHECK
#
#IF (DATE_CREATED GTE 20090101) config=(check/2009_checks.mch)
(start/sample_start.mcs)(constant/constant1.mcn)(status/sample_status.mcq)
#
#IF (USERNAME EQ JDOE) AND (MODEL_UNIT EQ INCH) config=(check/user_checks.mch)
(start/default_start.mcs)(constant/inch.mcn)(status/sample_status.mcq)
#
ELSE config=(check/default_checks.mch)(start/nostart.mcs)(constant/mm.mcn)
(status/sample_status.mcq)
#
IF (NOCHECK) NOCHECK
```

***Example condition.mcc File in Notepad***

Note that each statement is on one line, but due to the length of the options and so that the image could fit on the page, the lines have been word wrapped in Notepad.

In plain English, this is what the *condition.mcc* file that was shown is doing:

*If the system parameter that holds the value of the date the model was created is greater than or equal to January 1, 2009 use the following files:*

- *2009_checks.mch*
- *sample_start.mcs*
- *constant1.mcn*
- *sample_status.mcq*

*If the current user running Creo has the username JDOE and the model is using inches for its units, use the following files:*

- *user_checks.mch*
- *default_start.mcs*
- *inch.mcn*
- *sample_status.mcq*

*If none of the above conditions are not found to be true, use the following files:*

- *default_checks.mch*
- *nostart.mcs*
- *mm.mcn*
- *sample_status.mcq*

*Or if checks were not set to run, don't run any checks at all.*

Let's discuss the available operators and parameters for building a conditional statement.

## Operators

Operators are used to compare numerical, logical or string parameters and return a value based on whether the comparison is found to be true or false. In other words, they are the "things" that are used to set up the condition to be evaluated. They are listed here, followed by a description:

| Operator | Description |
|---|---|
| **EQ** | Equal to |
| **NEQ** | Not Equal to |
| **GT** | Greater than |
| **GTE** | Greater than or equal to |
| **LT** | Less than |
| **LTE** | Less than or equal to |

## Evaluation Value

The evaluation value is the last part of each conditional statement, and it is the information that a parameter will be evaluated against. For example:

- "EQ 6" means a parameter whose value is equal to 6
- "EQ Purchased" means a parameter whose value is equal to the string of characters "P" – "u" – "r" – "c" – "h" – "a" – "s" – "e" – "d"

Wildcard characters can be used to further broaden or narrow the evaluation criteria if the evaluation being performed on a string parameter. Think of a wildcard character as a "fill in the blank here" type of situation.

The wildcard characters and their meanings are:

| Character | Description |
|---|---|
| * | Any number of characters |
| ? | One character |
| # | One numerical character |
| $ | One string character |

Again, an example is worth a thousand words. Let's say we have the following information available to query in the model:

| Parameter Name | Parameter Type | Value |
|---|---|---|
| **USERNAME** | STRING | RASTHEIMER |
| **VOLUME** | REAL NUMBER | 2345 |
| **DATE_CREATED** | STRING | 20150624 |
| **CATEGORY** | STRING | P2 |
| **DEPARTMENT** | STRING | A3 |

The following table shows some example evaluation values and the names of the parameters that they would match if the evaluation were successful.

| Evaluation Value | Matching Parameter(s) |
|---|---|
| **2*** | VOLUME<br>DATE_CREATED |
| **$2** | P2 |
| **$#** | A3 |
| **201506##** | DATE_CREATED |
| **####** | VOLUME |
| **201501##** | <no match> |
| **###** | <no match> |

The remainder of this section gives a more detailed description of the available system parameters that might be relevant in a model-based environment.

## Parameters

A parameter in Creo is a piece of information, often referred to as metadata, which is connected to the model but isn't necessarily associated to geometry and therefore may not be visible. Parameters can be added and defined by the user, but several available system parameters are also automatically updated directly from the model. The remaining information in this section lists the relevant system parameters that are available to be checked and which conditional modifiers can be used to evaluate their value.

### DATE_CREATED

What: A system parameter that is set to the date that the model was created. Although no mechanism exists to view this information in the User Interface, you might be able to query for it if you are using a PDM system. Otherwise, you'll have to take a leap of faith that the information is there and you can reference its value using the format YYYYMMDD, where YYYY denotes year, MM denotes month and DD denotes day.

Can be used with operators: EQ, NEQ, GT, GTE, LT, LTE

Can be used with wildcard characters: * ? # &

Why: You may have checks that you do or don't want to run based on when the model was created. For example, let's say you started adding a parameter to models on March 21, 2012. You could run different checks on models that were created after that date.

Example Used in a Statement:

IF (**DATE_CREATED** LT 20120321) config = *<list of files>*

IF (**DATE_CREATED** GTE 20120321) config = *<list of files>*

## FT_GENERIC_ASSEMBLY

What: A system parameter that is used to indicate if the model is a **generic model** in a **family table assembly**. As with the other system parameters, no mechanism is available in the User Interface to view this parameter's value, but you could go to the Tools menu and launch the Family Table dialog to see if any family table design variations are defined.

Can be used with operators: <none>

Why: This check cannot be used with any operators because it simply evaluates to TRUE if the assembly contains instances, or FALSE if not.

Example Used in a Statement:

IF (**FT_GENERIC_ASSEMBLY**) config = *<list of files>*

## FT_GENERIC_PART

What: A system parameter that is used to indicate if the model is a **generic model** in a **family table part**. As with the other system parameters, no mechanism is available in the User Interface to view this parameter's value, but you could go to the Tools menu and launch the Family Table dialog to see if any family table design variations are defined.

Can be used with operators: <none>

Why: This check cannot be used with any operators because it simply evaluates to TRUE if the part contains instances, or FALSE if not.

Example Used in a Statement:

IF (**FT_GENERIC_PART**) config = *<list of files>*

## FT_INSTANCE_ASSEMBLY

What: A system parameter that is used to indicate if the model is an **instance** in a **family table assembly**. As with the other system parameters, no mechanism is available in the User Interface to view this parameter's value, but you could go to the Tools menu and launch the Family Table dialog to see if any family table design variations are defined.

Can be used with operators: <none>

Why: This check cannot be used with any operators because it simply evaluates to TRUE if the assembly is an instance, or FALSE if not.

Example Used in a Statement:

IF (**FT_INSTANCE_ASSEMBLY**) config = *<list of files>*

## FT_INSTANCE_PART

What: A system parameter that is used to indicate if the model is an **instance** in a **family table part**. As with the other system parameters, no mechanism is available in the User Interface to view this parameter's value but you could go to the Tools menu and launch the Family Table dialog to see if any family table design variations are defined.

Can be used with operators: <none>

Why: This check cannot be used with any operators because it simply evaluates to TRUE if the part is an instance, or FALSE if not.

Example Used in a Statement:

IF (**FT_INSTANCE_PART**) config = *<list of files>*

## GROUPNAME

What: The name of the group to which the current user belongs. To define a group, create a text file named <groupname>.mcg in the *modchk/text/usascii/config/groups* directory, where <groupname> is the name of the group. Each line in the file should contain the username of the user to belong to that group.

Can be used with operators: EQ, NEQ

Can be used with wildcard characters: * ? # &

Why: You may allow or require which checks users can run based on the group to which they belong.

Example Used in a Statement:

IF (GROUPNAME EQ Administrators) config = *<list of files>*

IF (GROUPNAME EQ Engineering) config = *<list of files>*

## MODEL_TYPE

What: A system parameter equal to one of the following values that describes the type of file that is being checked. Again, there is no way to view this information in the User Interface, and you might refer to the file extensions to at least determine if you are checking a part (.prt) or assembly (.asm), but many of the subtypes share the same extensions as their parent type:

PRT_SOLID

PRT_SHEETMETAL

PRT_SKELETON

PRT_PIPE

PRT_HARNESS

ASM_DESIGN

ASM_INTERCHANGE

ASM_MOLD_LAYOUT

Can be used with operators: EQ, NEQ

Can be used with wildcard characters: * ? # &

Why: You may want to check for different information based on the type of file that is being checked.

Comments: Note that no checks are available for drawings.

Example Used in a Statement:

IF (**MODEL_TYPE** EQ PRT_SOLID) config = *<list of files>*

IF (**MODEL_TYPE** EQ PRT_SHEETMETAL) config = *<list of files>*

## MODEL_NAME

What: Exactly as it sounds, this check will allow you to run a check against the name of the file being checked, minus the file extension such as .prt or .asm.

Can be used with operators: EQ, NEQ

Can be used with wildcard characters: * ? # &

Why: This check might be used to verify that the naming convention of the model adheres to standards that may have been set by your organization.

Example Used in a Statement:

IF (**MODEL_NAME** EQ 123-*) config = *<list of files>*

IF (**MODEL_NAME** EQ 456-*) config = *<list of files>*

## MODEL_UNIT

What: A system parameter that is used to set the units of the model. The units of the current file can be viewed in the User Interface by going to the File menu and selecting Prepare, and then Model Properties.

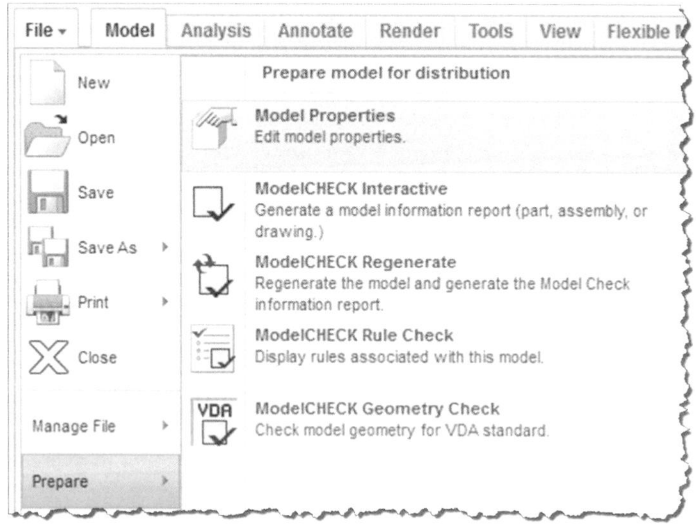


***How to Determine Units of a Model***

Possible values include: CM, FT, IN, M, MM

Can be used with operators: EQ, NEQ

Why: The use of different units might mean different tolerance for certain types of checks, so this check will allow you to customize what you check for by the units that have been set.

Comment: Be aware that even though FT, MICRON, MIL and MILE are listed as valid units in a Creo file, use of these units in a conditional statement causes a ModelCHECK error and result in the ModelCHECK report not being generated.

Example Used in a Statement:

IF (**MODEL_UNIT** EQ IN) config = *<list of files>*

IF (**MODEL_UNIT** EQ MM) config = *<list of files>*

## PRO_VERSION

What: A system parameter indicating the version of Creo (or Pro/ENGINEER) that was used when the model was last saved. The format of the value is: YYYYWW#, where YYYY and WW denote the year and week in which the version was released, and # is a single number value that represents the number of the build that week, starting with zero.

Can be used with operators: EQ, NEQ, GT, GTE, LT, LTE

Comment: This was the format used to identify the software version through Wildfire in 2004. Beginning in 2004, the format changed to the name of the version followed by "M#," where # was representative of a build number, which usually comes in increments of 10. For example, in Creo 3.0, go to the File menu, choose Help and then About PTC Creo. The dialog that displays will show:

*Release: 3.0*

*Date Code: M040*

***DOCUMENT CS187441 FROM PTC'S TECHNICAL SUPPORT WEBSITE CONFIRMS THAT THIS FUNCTIONALITY IS NOT SUPPORTED IN CREO PARAMETRIC.***

Why: This could be useful if what you are checking for varies based on the software version used. Perhaps some new functionality was introduced in a particular version for which you want to run a check. This evaluation criterion could be the deciding factor if you are setting up ModelCHECK to run in an older version of Pro/ENGINEER.

Example Used in a Statement:

IF (**PRO_VERSION** LT 2003042) config = *<list of files>*

## USERNAME

What: A system parameter set to the name of the user that is currently logged into the machine running Creo.

Can be used with operators: EQ, NEQ

Can be used with wildcard characters: * ? # &

Why: You may require certain checks to be run based on who is running ModelCHECK.

Comment: In my opinion, this is a bit of a strange filter. I guess the username could then determine the user's job function, which could then determine what checks would be run, but wouldn't this more easily be done through a group?

Example Used in a Statement:

IF (**USERNAME** EQ jdoe) config = *<list of files>*

IF (**USERNAME** EQ jdeer) config = *<list of files>*

## <PARAMETER>

What: Any parameters that have been defined in the file (model, assembly or drawing) can be referred to by their names. Note that parameter names are case sensitive.

Can be used with operators:

EQ, NEQ if the parameter is a string

EQ, NEQ, GT, GTE, LT, LTE if the parameter is a numeric value

Why: There's a good chance that you will want to check other information in your model, so here's your chance to customize your ModelCHECK installation!

Example Used in a Statement[xix]:

IF (**COMPANY** EQ Coca-Cola) config = <*list of files*>

IF (**COMPANY** EQ Pepsi) config = <*list of files*>

And here's another update to our visualization of the files being read after we've incorporated the *condition.mcc* file:

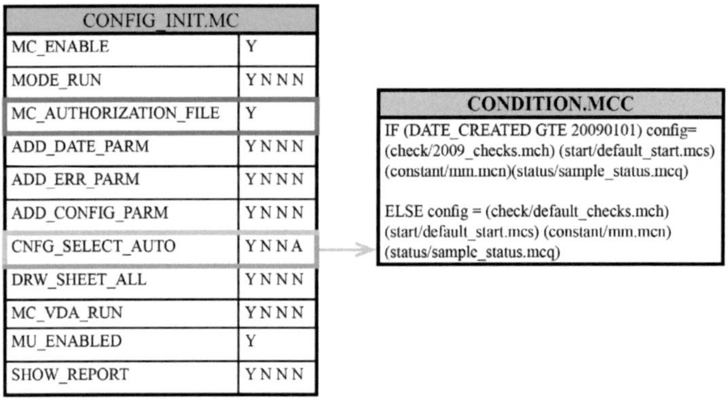

| CONFIG_INIT.MC | |
|---|---|
| MC_ENABLE | Y |
| MODE_RUN | Y N N N |
| MC_AUTHORIZATION_FILE | Y |
| ADD_DATE_PARM | Y N N N |
| ADD_ERR_PARM | Y N N N |
| ADD_CONFIG_PARM | Y N N N |
| CNFG_SELECT_AUTO | Y N N A |
| DRW_SHEET_ALL | Y N N N |
| MC_VDA_RUN | Y N N N |
| MU_ENABLED | Y |
| SHOW_REPORT | Y N N N |

| CONDITION.MCC |
|---|
| IF (DATE_CREATED GTE 20090101) config= (check/2009_checks.mch) (start/default_start.mcs) (constant/mm.mcn)(status/sample_status.mcq)<br><br>ELSE config = (check/default_checks.mch) (start/default_start.mcs) (constant/mm.mcn) (status/sample_status.mcq) |

***Mapping of ModelCHECK Files Including condition.mcc***

# SETCONF.MCC

The previous section discussed the file that would be read if the *CNFG_SELECT_AUTO* configuration option was set to Y in the config_init.mc file. But if you prefer that ModelCHECK not select a configuration based on a set of predetermined conditions, or that ModelCHECK asks the current user which configuration to apply, this is when this option is set to N or A, respectively, and this is when the *setconf.mcc* file is read.

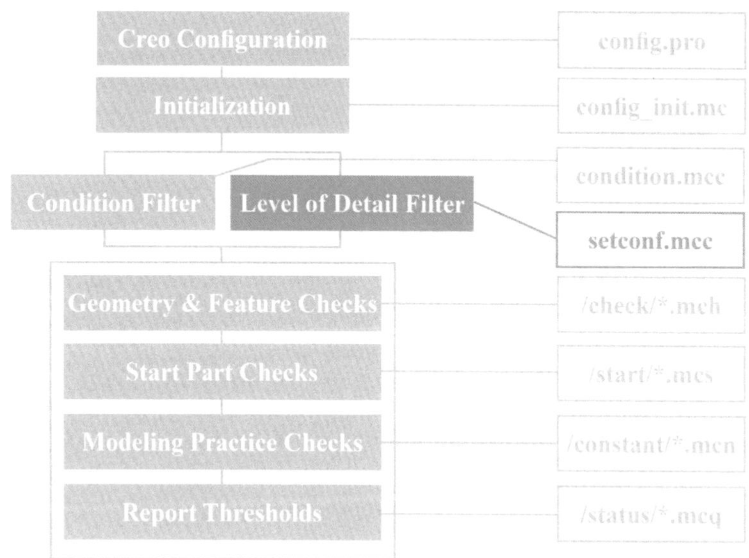


***Visual Representation of ModelCHECK Files***
***Focusing on the Level of Detail Filter File***

Why wouldn't you want to automatically select the configuration files? There are probably many different reasons, but a few possibilities might include:

- The logic of selecting the configuration files is too complex to be captured in a simple *IF* statement
- Several different configuration files could be selected based on the stage of development of the product, so allowing the user to select the configuration is simpler or more reliable.

- Perhaps you want to run more than one configuration, which could be done more easily by allowing the current user to select a configuration immediately before running ModelCHECK.

The *setconf.mcc* file is what determines the options that you will be presented with from the Creo "Load Config" menu in the user interface. But remember that the only time you will see a drop-down list in the User Interface is if the *CNFG_SELECT_AUTO* option is set to *N* or *A*. If the option is set to *Y*, this list doesn't even display in the user interface.

The Load Config drop-down list displays when you go to the File menu, choose Options and then select the Environment subsection. Right under the option to modify the ModelCHECK Settings you will see the "Load Config" option with a drop-down, as shown in the next figure.

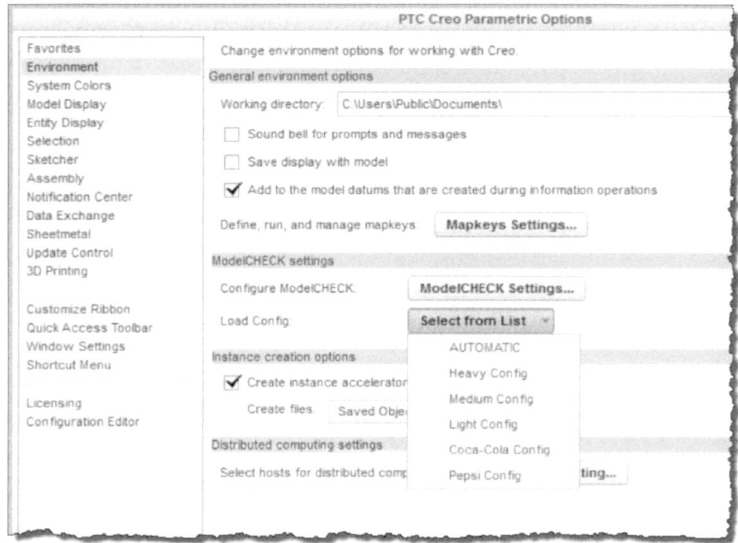


***Configuration Options List Available when the setconf.mcc File is Used***

The default ModelCHECK configuration of the *setconf.mcc* file contains three predefined levels of checking, Heavy, Medium and Light. More checking levels can be generated to build your own categories of checks to suit your organization. In the above example, you might work for Coca-Cola® and Pepsi® and each corporation requires different standards when model checking. In the dropdown menu after 'Load Config', I added two options named "Coca-Cola Config" and "Pepsi Config" to capture the different checking configurations.

The *setconf.mcc* file is a text file with a tag that identifies the labels, followed by a " = " and then the paths and names of the files to be read. The example file on the next page shows the two additional configurations added to the file in order to include the "Coca-Cola®" and "Pepsi®" options on the list.

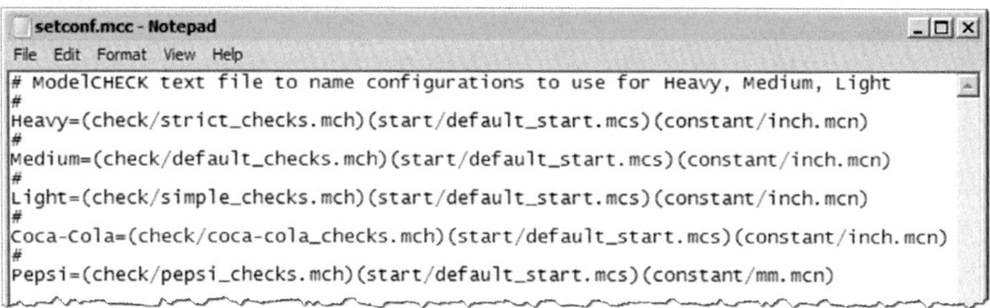

```
setconf.mcc - Notepad
File Edit Format View Help
# ModelCHECK text file to name configurations to use for Heavy, Medium, Light
#
Heavy=(check/strict_checks.mch)(start/default_start.mcs)(constant/inch.mcn)
#
Medium=(check/default_checks.mch)(start/default_start.mcs)(constant/inch.mcn)
#
Light=(check/simple_checks.mch)(start/default_start.mcs)(constant/inch.mcn)
#
Coca-Cola=(check/coca-cola_checks.mch)(start/default_start.mcs)(constant/inch.mcn)
#
Pepsi=(check/pepsi_checks.mch)(start/default_start.mcs)(constant/mm.mcn)
```

***Example setconf.mcc File in Notepad***

Another file named msg_mc.txt in the text directory is used to label the combinations presented to the user in the user interface. The syntax for these lines is %CILabel<#>, where # is a different number for each line in the file.

You can also include a tag that identifies a tooltip to be displayed when you mouse over the object. The syntax for these lines is %CIMessage<#>, where <#> corresponds to the number used in the label. The image below shows an example file that was used to set up the User Interface displayed in the previous image.

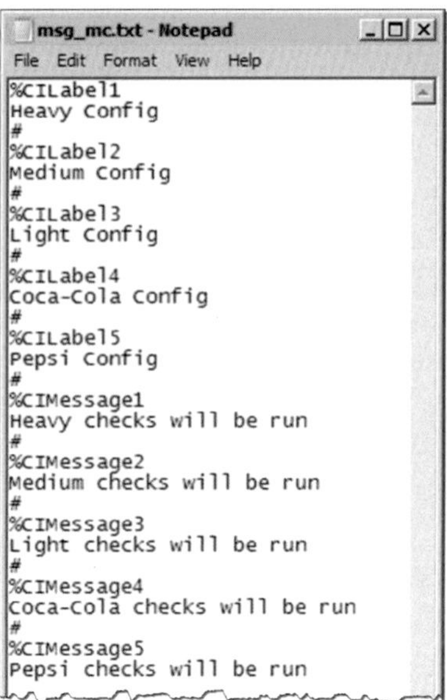


***Example msg_mc.txt File in Notepad***

If we take a look at the previous file, the following conditions are true:

- The 1st option in the drop-down list will be labeled "Heavy Config." Mousing over it, you will see a tooltip that says, "Heavy checks will be run."
- The 2nd option in the drop-down list will be labeled "Medium Config." Mousing over it, you will see a tooltip that says, "Medium checks will be run."
- The 3rd option in the drop-down list will be labeled "Light Config." Mousing over it, you will see a tooltip that says, "Light checks will be run."
- The 4th option will be "Coca-Cola Config." Mousing over it, you will see a tooltip that says, "Coca-Cola required checks will be run".
- The 5th option will be "Pepsi Config." Mousing over it, you will see a tooltip that says, "Pepsi required checks will be run."

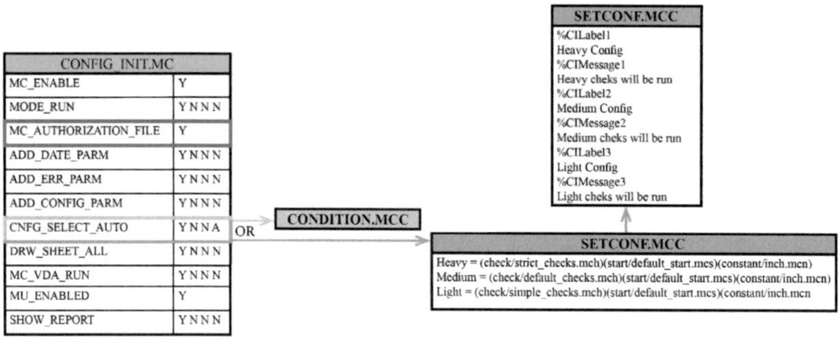

| CONFIG_INIT.MC | |
|---|---|
| MC_ENABLE | Y |
| MODE_RUN | Y N N N |
| MC_AUTHORIZATION_FILE | Y |
| ADD_DATE_PARM | Y N N N |
| ADD_ERR_PARM | Y N N N |
| ADD_CONFIG_PARM | Y N N N |
| CNFG_SELECT_AUTO | Y N N A |
| DRW_SHEET_ALL | Y N N N |
| MC_VDA_RUN | Y N N N |
| MU_ENABLED | Y |
| SHOW_REPORT | Y N N N |

OR

**CONDITION.MCC**

| SETCONF.MCC |
|---|
| %CILabel1 |
| Heavy Config |
| %CIMessage1 |
| Heavy cheks will be run |
| %CILabel2 |
| Medium Config |
| %CIMessage2 |
| Medium cheks will be run |
| %CILabel3 |
| Light Config |
| %CIMessage3 |
| Light cheks will be run |

| SETCONF.MCC |
|---|
| Heavy = (check/strict_checks.mch)(start/default_start.mcs)(constant/inch.mcn) |
| Medium = (check/default_checks.mch)(start/default_start.mcs)(constant/inch.mcn) |
| Light = (check/simple_checks.mch)(start/default_start.mcs)(constant/inch.mcn |

***Mapping of ModelCHECK Files Including setconf.mcc***

# *.MCH

The *\.mch* file is what I consider to be one, if not **the** most important, of the files in the setup of ModelCHECK. It contains checks for parts, assemblies and drawings and is the connection to many other files that can check the integrity of your geometry, amongst other things.

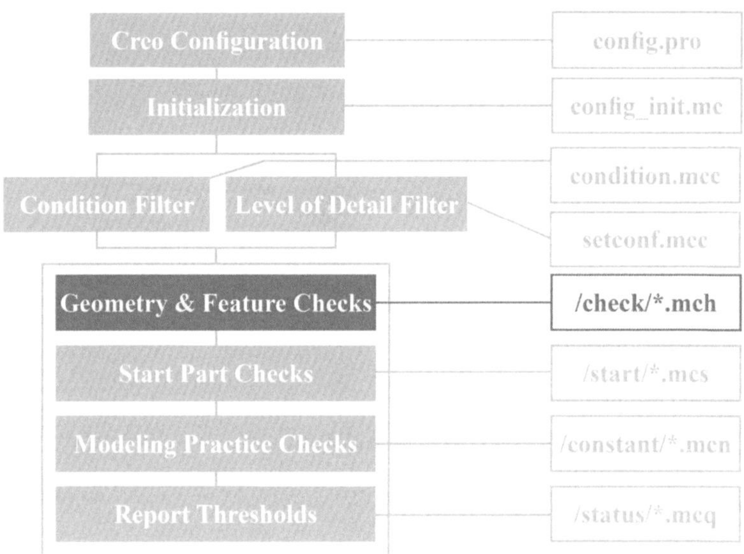


***Visual Representation of ModelCHECK Files***
***Focusing on the Geometry & Feature Checks File***

The previous two files that we just covered (*condition.mcc* and *setconf.mcc*) will either evaluate the condition that you've set (in the *condition.mcc* file) or the option that you've selected from the drop-down list in the User Interface (set by the *setconf.mcc* file) and are what determines which *\.mch* file is read.

The syntax of this file has the same format that we saw in the first ModelCHECK file that we reviewed (config_init.mc). It starts with the option followed by the allowable values and then four letters that represent the setting for that option in Interactive, Batch, Regenerate and Save modes. The difference here is that these checks can be set to flag the results as an error or warning. This helps delineate between conditions that are critical and not, by flagging them with an error or warning respectively.

```
default_checks.mch - Notepad
File  Edit  Format  View  Help
AE_GTOL_DUPLICATE      YNEW   Y   N   N   N
AE_MISSINGREFS         YNEW   Y   N   N   N
AE_SF_DUPLICATE        YNEW   Y   N   N   N
AF_INCOMPLETE          YNEW   Y   N   N   N
ANNTN_INACTIVE         YNEW   Y   N   N   N
CYL_CUT_SLOTS          YNEW   Y   N   N   N
CYL_DIAMS              YNEW   Y   N   N   N
EARLY_CHAMFER          YNEW   W   N   N   N
EARLY_COSMETIC         YNEW   W   N   N   N
EARLY_DRAFT            YNEW   W   N   N   N
EARLY_ROUND            YNEW   W   N   N   N
EDGE_REFERENCES        YNEW   W   N   N   N
```

***Example *.mch File in Notepad***

Taking a look at the file above, you will notice that five modes are listed to be set. The four that I mentioned previously (I, B, R, S) are still valid, but you will also see "M." M stands for metrics and was formerly used to collect metrics on the ModelCHECK data that was a result of the checks. If you look at more recent versions of the documentation, this option is no longer listed. In fact, the last time it was listed required Internet Explorer version 6.0, released in 2001, and it may still be lingering in some of the files.

Over 100 options can be set in this file. Some are applicable to parts, some to assemblies, and some to drawings. A note under each option indicates at what level the check can be run. I won't cover all of the available checks in this book (that's what the documentation is for!), but I will cover the ones that are relevant given the best practices outlined in Re-Use Your CAD: The Model-Based CAD Handbook, 1st Edition (and a few others that are just plain good ol' modeling practice), to help ensure you are adhering to recommended practices.

These options will follow the same format we've used for other files:

## OPTION_NAME <VALUES> SETTINGS FOR "I B R S"

What: Description of what the option does.

Why: Why you might want to use or not use this option.

Where: This will explain where the option can be used.

Comments: Further comments.

## AE_GTOL_DUPLICATE <YNEW> Y N N N

What: This will check for **duplicate** Annotation Elements within the model. Elements are considered duplicate if they have the same references and are of the same type. Any additional options or modifications that you have specified on the geometric tolerance must also match. Conditions flagged as an error when the value is set to "E," or a warning when the value is set to "W."

Why: On the odd chance that more than one person captured the same geometric tolerance information in the same model (perhaps one of them ended up in a combination state, was not visible, and then was unintentionally duplicated), ModelCHECK will automatically detect this situation for you.

Where: Part and Assembly

Comments: Remember that an Annotation Element is one feature that is part of an Annotation Feature. The left image below highlights one Annotation Element inside of an Annotation Feature. The right image below shows the Annotation Feature dialog that lists the four defined Annotation Elements.

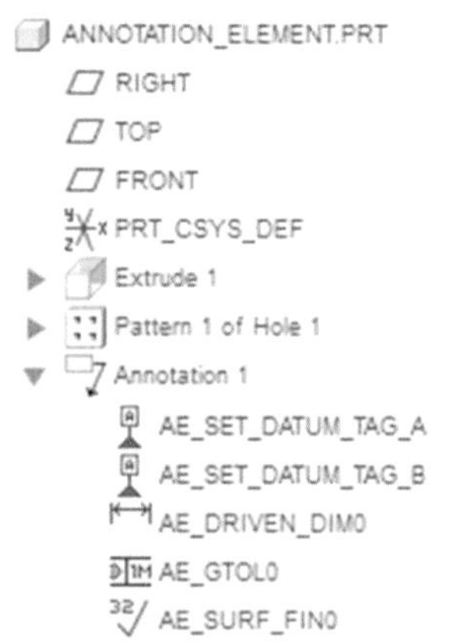

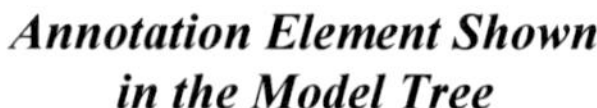

***Annotation Element Shown in the Model Tree***

***Annotation Feature Containing Four Annotation Elements***

## AE_MISSINGREFS <YNEW> Y N N N

What: This will report all Annotation Elements that have **missing references**.

Why: This is typically a result of a feature that has been deleted and was being referenced in a geometric tolerance. Geometric tolerances that are left “hanging in the breeze” won’t do you, or any downstream application that may need to read that data, any good, so this is a good sanity check to make sure that all of the bases are covered.

Where: Part

Comments: Remember that an Annotation Element is one feature that is part of an Annotation Feature. Refer to the image in the option AE_GTOL_DUPLICATE.

## AE_SF_DUPLICATE <YNEW> Y N N N

What: This will check all **surface finish** Annotation Elements to ensure that none of them have any surface references in common.

Why: While having more than one surface finish annotation in a model might be acceptable, you will want to ensure that you don’t have more than one surface finish definition referencing the same surface. This could result in conflicting information and might make it impossible to know which definition is correct.

Where: Part and Assembly

Comments: Remember that an Annotation Element is one feature that is part of an Annotation Feature. Refer to the image in the option AE_GTOL_DUPLICATE

## AF_INCOMPLETE <YNEW> Y N N N

What: If an Annotation Feature contains at least one Annotation Element that is missing strong references, this check will be flagged.

A strong reference for an Annotation Feature is defined when an Annotation Element is attached to an annotation plane, entities of a geometric tolerance, a driven dimension or a reference dimension.

Why: Perhaps you might need to leave an Annotation Element incomplete or with missing references because they aren't available when you are defining the annotation, but you want to make sure that all information has been captured before releasing the model for review. This is a good way to double-check that you haven't missed anything and that all loose ends have been tied up.

Where: Part and Assembly

## ANNTN_INACTIVE <YNEW> W N N N

What: Reports any **inactive Annotation Elements** in the model.

Why: If you leave an Annotation Element definition incomplete due to missing information or geometry, or if an Annotation Element becomes incomplete due to a change in geometry, you will want to make sure that this element is up-to-date before releasing the model.

Where: Part and Assembly

Comments: The global reference viewer, which can be accessed from the ModelCHECK report, will also show inactive Annotation Elements.

## CYL_CUT_SLOTS <YNEW> E N N N

What: This check will look for any features in the model that use a sketch of a circle or slot. This is in an attempt to identify holes that were not created as a hole feature, as discussed on page 111 in the print version of Re-Use Your CAD: The Model-Based CAD Handbook, 1st Edition.

Why: On the surface (pun intended!), the geometry will appear identical, but the hole definition contains hidden metadata that might be used downstream for manufacturing or inspection purposes. If your goal is to achieve Model-Based Definition, this is a rule that must be followed and is a fairly easy thing to check.

Where: Part

Comments: This is one of my favorite examples of "incorrect modeling" that I present to my undergraduate students because it is easy to understand. Given that CAD modeling started long before Model-Based Definition was conceived, it is feasible to understand why a designer might have used an extruded sketch to create a hole. The resulting geometry was the same, so what was the harm in creating it as an extruded sketch? In past versions, there was no harm and, therefore, such features could continue to linger in legacy models.

***IF A HOLE WAS CREATED BY REVOLVING A RECTANGULAR SKETCH, IT WILL NOT BE DETECTED BY THIS CHECK.***

***The Geometry of the Three Holes in this Model Appear to be Identical, but the Hole Created by Revolving a Sketch will not be Detected by ModelCHECK***

## CYL_DIAMS <YNEW> Y N N N

What: This will look at all circular cuts in the model to ensure that they use a standard diameter value. The allowable diameter values are stored in an external text file named STD_HOLE_DIAM, that is pointed to in the next file that we will cover, the **.mcs* file. Any circular cuts in the model that do not have a standard diameter from this list will be flagged.

Why: This check helps ensure that standard parts are used as often as possible, as recommended on page 111 in the printed version of <u>Re-Use Your CAD: The Model-Based CAD Handbook, 1st Edition</u>. Another possibility is that the facility manufacturing the product you are designing has a limited number of diameters that can be drilled. ModelCHECK can be set up to help ensure that your designers have not designed something that cannot be manufactured "as desired." I say "as desired" because, if you couldn't drill the hole, you could probably 3D print it.

Where: Part

## EARLY_CHAMFER <YNEW> Y N N N

What: This will report **chamfers** that appear "early" in the model tree. The percentage value can be set as an option named PERC_EARLY_CHAMF in the **.mcn* file (which we will discuss in a future section).

Why: Generally, the stability of a model is better when finishing features, such as chamfers, are added to the model toward the end of the design process. This is covered on pages 109-110 in the printed version of Re-Use Your CAD: The Model-Based CAD Handbook, 1st Edition.

Where: Part

Comments: It isn't obvious that the default datum features, such as datum planes and the Csys, are included in the model tree count. Why does this matter? It will slightly change the tipping point. Let's use an example to explain.

Let's say you have a model with four features with the check set to 0.50 (refers to 50%, which we will cover when we discuss the **.mcn* file options). This means that the early chamfer check will look for chamfers in the first half of the model tree.

You run the check, but chamfer isn't flagged as being "early" in the model. Why not? It is the second feature out of four, so it should have been flagged!

Once you know that the default datums are included in the feature count, you can see why you'd have to set the percentage to 0.75 in order to consider the chamfer an early feature.

| Feature | Feature No. | % Position in Model Tree |
|---|---|---|
| CSYS | 1 | < 0.125 |
| DTM TOP | 2 | < 0.25 |
| DTM FRONT | 3 | < 0.375 |
| DTM RIGHT | 4 | < 0.5 |
| EXTRUDE | 5 | < 0.625 |
| CHAMFER | 6 | < 0.75 |
| HOLE | 7 | < 0.875 |
| PATTERN | 8 | < 1 |

***Datum Features are Included when Identifying Features in the Model Tree***

So what is the magic number that should be set? This is hard to say because it really depends on the complexity of your models.

One approach would be to take a look at a random sample of files and count the number of features in each file. This would allow you to come up with an average number of features per model so that you can make an educated guess of what “early” means. Of course, the more features in a model, the less impact the default datums will have on the check being flagged.

The other approach would be to start at 0.50 and see how it works for you. You should manually check for early chamfers at first and, if you notice that the check is not picking up what you consider to be early, it's a good indication that the value needs to be larger.

## EARLY_COSMETIC <YNEW> Y N N N

What: This will report **cosmetic** features that appear "early" in the model tree. The percentage value can be set as an option named PERC_EARLY_COSMETIC in the **.mcn* file (which we will discuss in a future section).

Why: Generally, the stability of a model is better when finishing features such as cosmetic threads are not explicitly modeled and are added to the model toward the end of the design process. This is covered on page 111 - 113 in the print version of Re-Use Your CAD: The Model-Based CAD Handbook, 1st Edition.

Where: Part

Comments: Refer to the explanation found in the EARLY_CHAMFER check above as the same principles apply.

## EARLY_DRAFT <YNEW> Y N N N

What: This will report **draft** features that appear "early" in the model tree. The percentage value can be set as an option named PERC_EARLY_DRAFT in the **.mcn* file (which we will discuss in a future section).

Why: Generally, the stability of a model is better when finishing features such as drafts are added to the model toward the end of the design process. This is covered on pages 109-110 in the print version of Re-Use Your CAD: The Model-Based CAD Handbook, 1st Edition.

Where: Part

Comments: Refer to the explanation found in the EARLY_CHAMFER check above as the same principles apply.

## EARLY_ROUND <YNEW> Y N N N

What: This will report **round** features that appear "early" in the model tree. The percentage value can be set as an option named PERC_EARLY_ROUND in the **.mcn* file (which we will discuss in a future section).

Why: Generally, the stability of a model is better when finishing features such as a round feature are added to the model toward the end of the design process. This is covered on pages 109-110 in the print version of Re-Use Your CAD: The Model-Based CAD Handbook, 1st Edition, but if you read the previous three options and explanations, you've probably figured that out by now.

Where: Part

Comments: Refer to the explanation found in the EARLY_CHAMFER check above, as the same principles apply.

## EDGE_REFERENCES <YNEW> W N N N

What: This will check features in the model defined using dimensions referring to an **edge** in the model.

Why: This used to be a bigger deal in older versions of the software, in my opinion. Let's say you reference an edge to locate a feature, and then that edge is rounded later on in the design process. Technically, that edge no longer exists in the model and, therefore, might cause a regeneration error. I've found that this particular example usually isn't a problem because PTC has figured out how to derive the location of the edge and reattach the dimensions to this theoretical intersection (behind the scenes, without you even knowing it).

I'm sure *some* situations probably exist that wouldn't be automatically dealt with, but it can't hurt to turn on this check. This check could be thought of as a way to make sure design intent was captured in an intentional and thoughtful way.

Where: Part

## OPTIONS RELATED TO MC_VDA

Of the 30 options that can be set relating to the GeomIntegrityCheck turned on in the config_init.mc file, two in particular should be checked when taking a model-based approach.

### S026_MULT_BODY <EW> E

What: If components in an assembly contain more than one solid volume, this condition will be flagged as an error when the value is set to "E," or a warning when the value is set to "W."

Why: Page 83 of Re-Use Your CAD: The Model-Based CAD Handbook, 1st Edition suggests having only one solid body in a model, which is consistent with every CAD system I've ever come across.

### S027_MULT_SOLID <EW> E

What: If two solid bodies do not touch each other, but are in the same model, this condition will be flagged as an error when the value is set to "E," or a warning when the value is set to "W."

Why: This is almost a duplicate check of the previous option, but it looks for this condition in the context of a model, rather than in the context of an assembly. This condition may be necessary for 3D printing in cases where multiple components are unified in a single model for build nesting, or for creating an assembly in a one-shot print.

Another update to our visualization of the files.

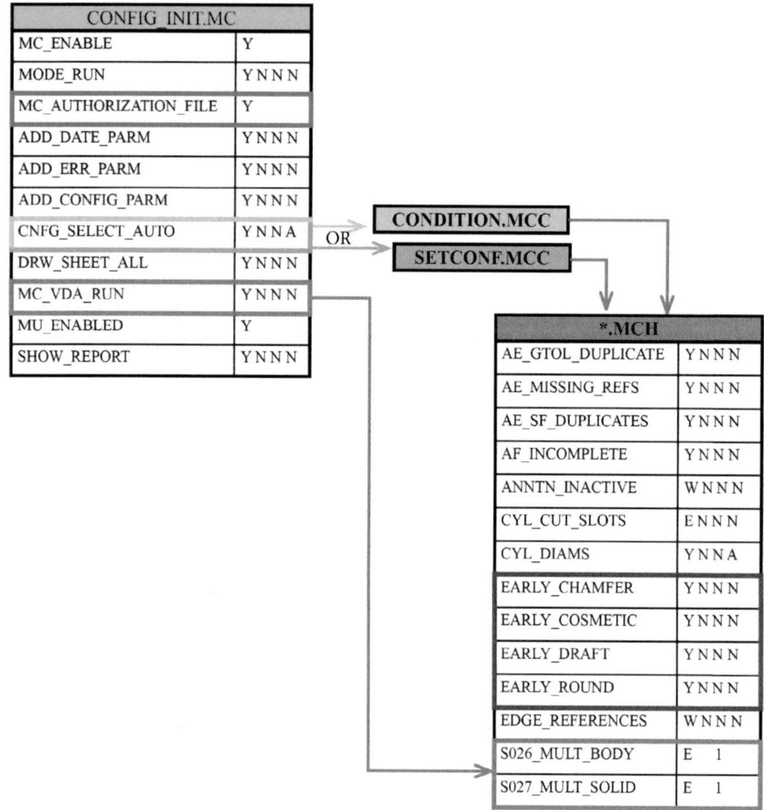


**Mapping of ModelCHECK files including *.*mch***

# *.MCS

The options found in this file are ones that might benefit anyone. Whether a model-based approach is what you are after or not, this file will help ensure that the content in your models is consistent. If you use start parts (a practice that is quite common and considered good practice), you are less likely to need to rely on these checks, but as I've said several times already – the checks are available for your use, so not taking advantage of them seems silly.

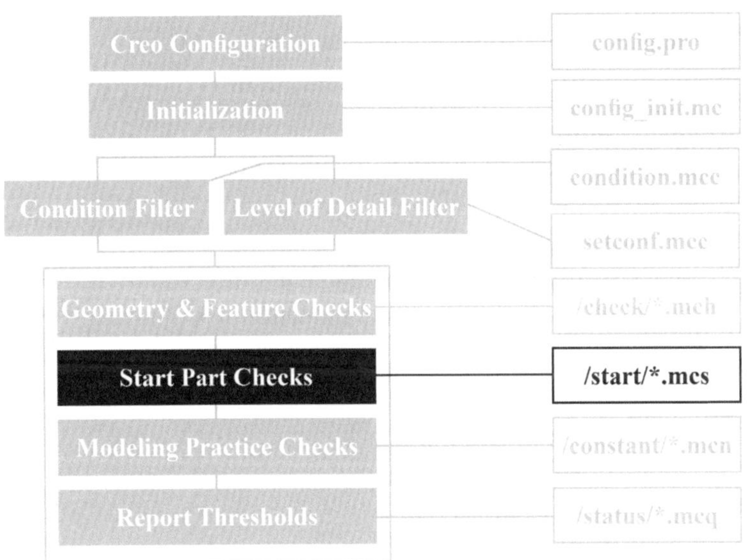


***Visual Representation of ModelCHECK Files***
***Focusing on the Start Part Checks File***

A standard set of options can be put into this file, but the values for which they are checking can be set to anything. In other words, this file will allow you to check for standard features, such as datum planes and parameters, but the value to which the options will be set will depend on what you've decided to name them.

## Standard Datum Features

What: Several checks will allow you to check for standard datum features such as axes, coordinate systems, curves, planes and points in either a model or assembly. You can also check their position in the model tree.

The formula for setting up this check is:

<A>_DATUM_<B> <NAME> <# IN MODEL TREE>

where

<A> = ASM or PRT

<B> = AXIS, CSYS, CURVE, PLANE or POINT

<NAME> = the name of the object from the list of <B>

<# IN MODEL TREE> = a single integer that lists where this datum is expected to be found in the model tree.

Possible combinations for the options are:

ASM_DATUM_AXIS <# IN MODEL TREE>
PRT_DATUM_AXIS <# IN MODEL TREE>
ASM_DATUM_CSYS <# IN MODEL TREE>
PRT_DATUM_CSYS <# IN MODEL TREE>
ASM_DATUM_CURVE <# IN MODEL TREE>
PRT_DATUM_CURVE <# IN MODEL TREE>
ASM_DATUM_PLANE <# IN MODEL TREE>
PRT_DATUM_PLANE <# IN MODEL TREE>
ASM_DATUM_POINT <# IN MODEL TREE>
PRT_DATUM_POINT <# IN MODEL TREE>

Why: These are all good checks to ensure that either a start model was used that contains your company standard for default datums, or that a standard procedure to set up default datums in the model was followed. Pages 92 – 93 of Re-Use Your CAD: The Model-Based CAD Handbook, 1st Edition discusses the use of Default References and Annotation Planes, which could be setup in your start model.

Where: Part, if the option begins with PRT.

Assembly, if the option begins with ASM.

Comments: Based on the recommended setup shown in the image on page 93 in the print version of Re-Use Your CAD: The Model-Based CAD Handbook, 1st Edition, the file would need to contain the following lines, with one exception. Spaces in the option names from Re-Use Your CAD: The Model-Based CAD Handbook, 1st Edition have been replaced with an underscore since Creo does not allow spaces when naming features. I've also shown an example of what the model tree for this start part might look like, including a column that shows the feature number, so you can see how it related to the value in the *.*mcs* file.

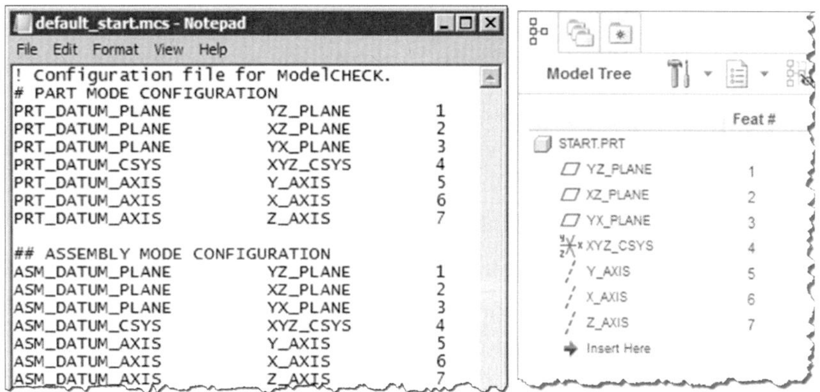

```
! Configuration file for ModelCHECK.
# PART MODE CONFIGURATION
PRT_DATUM_PLANE          YZ_PLANE      1
PRT_DATUM_PLANE          XZ_PLANE      2
PRT_DATUM_PLANE          YX_PLANE      3
PRT_DATUM_CSYS           XYZ_CSYS      4
PRT_DATUM_AXIS           Y_AXIS        5
PRT_DATUM_AXIS           X_AXIS        6
PRT_DATUM_AXIS           Z_AXIS        7

## ASSEMBLY MODE CONFIGURATION
ASM_DATUM_PLANE          YZ_PLANE      1
ASM_DATUM_PLANE          XZ_PLANE      2
ASM_DATUM_PLANE          YX_PLANE      3
ASM_DATUM_CSYS           XYZ_CSYS      4
ASM_DATUM_AXIS           Y_AXIS        5
ASM_DATUM_AXIS           X_AXIS        6
ASM_DATUM_AXIS           Z_AXIS        7
```



***Example *.mcs File in Notepad***

***Example Model Tree Showing Expected Layout of Model from a Start Part***

You may have other lines that you'd like to add based on your own company standards. For example, the Re-Use Your CAD: The Model-Based CAD Handbook, 1st Edition doesn't list default points or curves, but this might be a requirement if you are designing something that must connect at a certain location or has wires routed through it that are represented using curves.

## Annotation Planes & Views

What: This check will confirm the existence of annotation planes and views in the model. Page 94 in the print version of Re-Use Your CAD: The Model-Based CAD Handbook, 1st Edition recommends three default planes named Front, Right and Top that may be, but are not required to be, coincident with the default planes. An additional plane is recommended to hold notes.

The options that can be set are:

ASM_PLANE *<PLANE NAME>*

PRT_PLANE *<PLANE NAME>*

ASM_VIEW *<VIEW NAME>*

PRT_VIEW *<VIEW NAME>*

The option is then followed by the name of the plane or the saved view orientation.

For example, this check would look for a plane in an assembly labeled 'front' and a saved view named 'front'.

ASM_PLANE front

ASM_VIEW front

Why: Having specifically named orientations may be a requirement for a Technical Data Package (TDP) or a standard that your organization chooses to follow.

Where: Part, if the option begins with PRT.

Assembly, if the option begins with ASM.

# Metadata

What: Any data that needs to be captured but does not relate to a specific feature or part of the geometry can be captured in the model as a parameter. These pieces of information will be stored in Creo parts or assemblies as parameters, and the options to check for parameters in an assembly and part file respectively are:

ASM_PARAMETER
PRT_PARAMETER

These options can be used to check for the existence of the parameter as well as its parameter type and value. The general syntax is:

*<A>*_PARAMETER *<name>* *<type>* *<equal>* *<value>*

where:

*<A>* = ASM or PRT

*<name>* = parameter name

*<type>* = "STR" for a string
= "BOOL" for a boolen (yes/no)
= "INTEGER" for an integer value
= "REAL" for a real value
= "NONE" for a parameter that has no type set

*<equal>* = EQ, NEQ, LT, GT, LTE or GTE as discussed in detail on page 50

*<value>* = parameter value which can include wildcard characters as discussed on page 51.

| Example Check for Parameters: | Explanation |
|---|---|
| **PRT_PARAMETER MASS REAL** | check for a part parameter named MASS of type REAL |
| **PRT_PARAMETER COMPANY STR EQ ACTION*** | check for a part parameter named COMPANY of type STRING that starts with "ACTION" |
| **PRT_PARAMETER PHASE INT EQ ##** | check for a part parameter named PHASE and is of type INTEGER and has no more than two digits in its value |

Why: Pages 84 & 85 of the print version of Re-Use Your CAD: The Model-Based CAD Handbook, 1st Edition list metadata elements that should be uniform throughout a company for parts, assemblies, drawings, supplemental TDP documents, PDM and PLM. The exact names and descriptions may vary for each organization but as long as they are consistently used you will benefit from them.

Where: Part, if the option begins with PRT.

Assembly, if the option begins with ASM.

| Metadata Label (Parameter Name) | Description of Label |
|---|---|
| **Number** | Unique identifying number applied to each document. |
| **Description** | Text applied to each document that describes the data file to which it is attached. |
| **Revision** | Appropriate revision identifier per company release process.<br>For examples:<br>• A, B, C Letter Series<br>• WIP: in development prior to release stage<br>• C-WIP: Revision C in process of revision to Revision D. |
| **Revision_Date** | Date release occurred. |
| **Units** | Identify Metric or U.S. Customary |
| **NHA** | Identify all Next Higher Assemblies (NHA) to which this data set feeds. This may also be referred to by configuration management as "effectivity." |
| **Material** | Material identification reflected from the material library. |
| **Company** | Name of the responsible company. |
| **CAGE_Code** | Company cage code per Federal Cataloging Handbook H4/H8. |
| **Project_Number** | Contract number or identification. |
| **Originator_Name** | Name of the individual who created the document. |
| **Approver_Name** | Name of the individual who approves the file for release. |
| **Approver_Function** | Description of the function of the approver. |
| **Approval_Date** | Date of the approval. |
| **TDP_Type** | Identifies whether this data set is Model & Drawing or Model-Only. |
| **Data Rights** | Identify the appropriate requirements for restriction access, availability, proprietary data, or use of this digital file. |
| **Classification** | Identifies whether the file is UNCLASSIFIED or CLASSIFIED, with nomenclature per DOD Manual 5220.22-M. |

***Recommended Parameters to Check For Based on <u>Re-Use Your CAD: The Model-Based CAD Handbook, 1st Edition</u>***

Why: Having specifically named metadata may be a requirement for your organization or for a Technical Data Package (TDP) if you are adhering to a specific standard such as MIL-STD-31000A, discussed on page 84 - 85 of the print version of Re-Use Your CAD: The Model-Based CAD Handbook, 1st Edition.

Where: Part, if the option begins with PRT.

Assembly, if the option begins with ASM.

Note: Spaces in the option names from Re-Use Your CAD: The Model-Based CAD Handbook, 1st Edition have been replaced with an underscore since Creo does not allow spaces in a parameter name.

## Tolerances

What: Page 89 of the print version Re-Use Your CAD: The Model-Based CAD Handbook, 1st Edition suggests using a standard note block to define tolerances. While this check won't review title block content (this can be done through a check for the use of standard notes, which we'll get to), it will check to see the type of tolerance that are set up in the part or assembly file.

The options are:

ASM_TOL_TYPE

PRT_TOL_TYPE

and they can be set to either: DIN/ISO or ANSI

Why: This is another sanity check to make sure that the tolerances have been set up as you expect and have not unintentionally been changed.

Where: Part, if the option begins with PRT.

Assembly, if the option begins with ASM.

## File Naming and/or Numbering

What: As discussed on page 86 of the print version of Re-Use Your CAD: The Model-Based CAD Handbook, 1st Edition, either your company or the organization to which you will be providing data probably has a specific format for file naming.

Setting up this check is much like setting up the *condition.mcc* file, in that it allows you to completely customize what is being checked. And as you might have come to expect by now, based on this customizability, I can't recommend a standard set of values for this file, so I will explain how to set up the file to meet your needs.

Each line in the file will follow this formula:

<Mode>_<Type>_Name <Operator> <Evaluation Value>

Where:

**Mode**: This describes the type of file that will be checked. Possible values are:

ASM - Assembly

DRW - Drawing

PRT - Part

**Type**: This describes the status of the file. Possible values are:

Model - model

Instance - name of an instance of a family table part

Simprep - name of a Simplified Representation

**Operator:** Operators are used to compare numerical, logical or string parameters and return a value based on whether the comparison is found to be true or false. In other words, they are the "things" used to set up the condition to be evaluated. Operators are listed in the following table.

| Operator | Description |
|---|---|
| **EQ** | Equal to |
| **NEQ** | Not Equal to |
| **GT** | Greater than |
| **GTE** | Greater than or equal to |
| **LT** | Less than |
| **LTE** | Less than or equal to |

**Evaluation Value:** The evaluation value is the last part of each line, and it contains the information that will be looked for in the file name.

Wildcard characters can be used to further broaden or narrow the evaluation criteria on string parameters. Think of a wildcard character as a "fill in the blank" situation.

The wildcard characters and their meanings are:

| Character | Description |
|---|---|
| * | Any number of characters |
| **?** | One character |
| # | One numerical character |
| **$** | One string character |

Again, an example is worth a thousand words. The following shows example lines and explains the criteria to be checked.

| Evaluation Value | Evaluation Explanation |
|---|---|
| PART_INST_NAME EQ 001-* | In a part file, look for a family table instance starting with "001-" |
| ASM_MODEL_NAME EQ *_33331 | In an assembly file, look for "_33331" as the last 6 characters in the assembly's name |
| PART_SIMPREP_NAME EQ $#* | In a part file, look for a simplified representation that begins with a single character, followed by a single numerical character, followed by any additional characters. |
| PART_MODEL_NAME EQ 00?* | 001-6789.prt<br>and<br>003-pin.prt |

## Default Units

What: Page 83 of the print version of Re-Use Your CAD: The Model-Based CAD Handbook, 1st Edition talks about having a standard set of units. However, we know that units might vary based on your downstream users, so this is a great case to have a *condition.mcc* file in place to determine which **.mcs* file would be used, which will in turn determine the units to be checked.

The checks that can be used are named:

ASM_UNITS_LENGTH

PRT_UNITS_LENGTH

And can be set to:

INCH

MM

And named:

ASM_UNITS_MASS

PRT_UNITS_MASS

And can be set to:

KILOGRAM

POUND

Why: This is another sanity check to make sure that the units and mass have been set up as you expect and have not unintentionally been changed.

Where: Part, if the option begins with PRT.

Assembly, if the option begins with ASM.

## USE OF STANDARD NOTES

What: Notes are generally unique to each company, and as discussed on page 90 of the print version of Re-Use Your CAD: The Model-Based CAD Handbook, 1st Edition, by having a common note library that is accessible to all users, you can save time and maintain consistency.

Three different options in the *.*mcs* file allow you to:

- Check that existing parameter notes are from a standard library
- Check that required parameter notes exist in the file
- Check that unacceptable parameter notes do not exist in the file

The options are:

ASM_STD_NOTE

PRT_STD_NOTE

ASM_PARAM_NOTE_REQ

PRT_PARAM_NOTE_REQ

ASM_PARAM_NOTE_UNACC

PRT_PARAM_NOTE_UNACC

The options are then followed by the word PARAMETER and then either:

- The name of a text file (that exists in the config/text directory) containing a list of notes, or
- A string surrounded by double quotes.

For example, the following line:

*PRT_PARAM_NOTE_REQ STANDARD companyA_notes.txt*

would look for a parameter named STANDARD in a part file and would then check to ensure it was set to a value listed on one of the lines of a text file named companyA_notes.

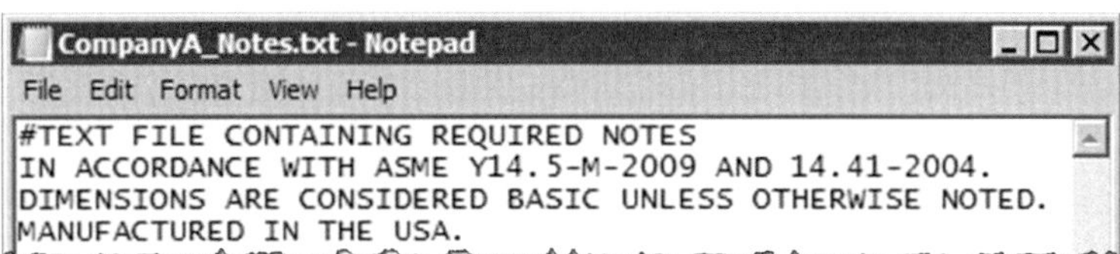


***Example Company Notes File in Notepad***

This line:

*ASM_PARAM_NOTE_UNACC STATUS "HOLD"*

would look for an assembly level parameter named STATUS and would flag an error if the value were set to HOLD.

Why: This check will help to ensure that notes are consistently used across an organization.

Where: Part, if the option begins with PRT.

Assembly, if the option begins with ASM.

## STANDARD SIZED HOLES

What: Having a standard hole library can also save time and ensure consistency, although for a slightly different reason discussed in the Why paragraph below.

The option in the **.mcs* file is STD_HOLE_DIAM_FILE, followed by the path to the file containing the standard hole sizes. The default name of the file is holes.txt, found in the mc\config\text directory.

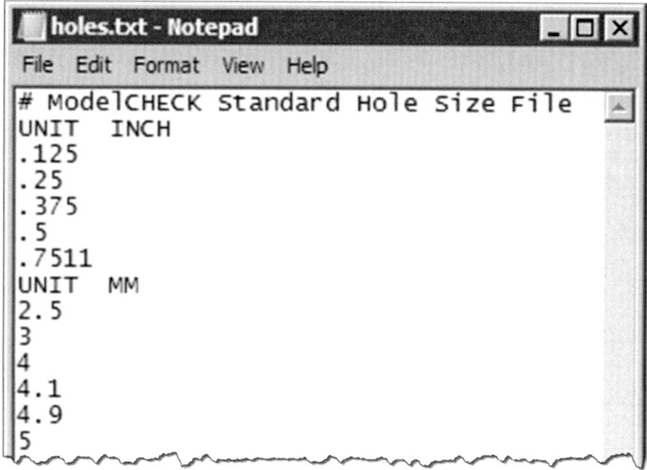

```
# ModelCHECK Standard Hole Size File
UNIT   INCH
.125
.25
.375
.5
.7511
UNIT   MM
2.5
3
4
4.1
4.9
5
```

***Example Standard Hole File in Notepad***

In order for the holes.txt file to be read, the option CYL_DIAMS must be set in the **.mch* file, as discussed earlier.

Why: You might use a standard hole library that contains only the hole sizes that you know can be manufactured given the facilities available for use. Or perhaps experience has taught you that making a hole slightly larger will have a large impact on the final cost of the product.

Where: Part files only.

As our mapping of the files used gets larger, I've chosen to omit the files that we've already covered on the left side of the image. Also, the *.*mcs* file here only shows checks that will be applied to parts but, as the previous options explained, some of these checks can also be set up to check assembly files as well.

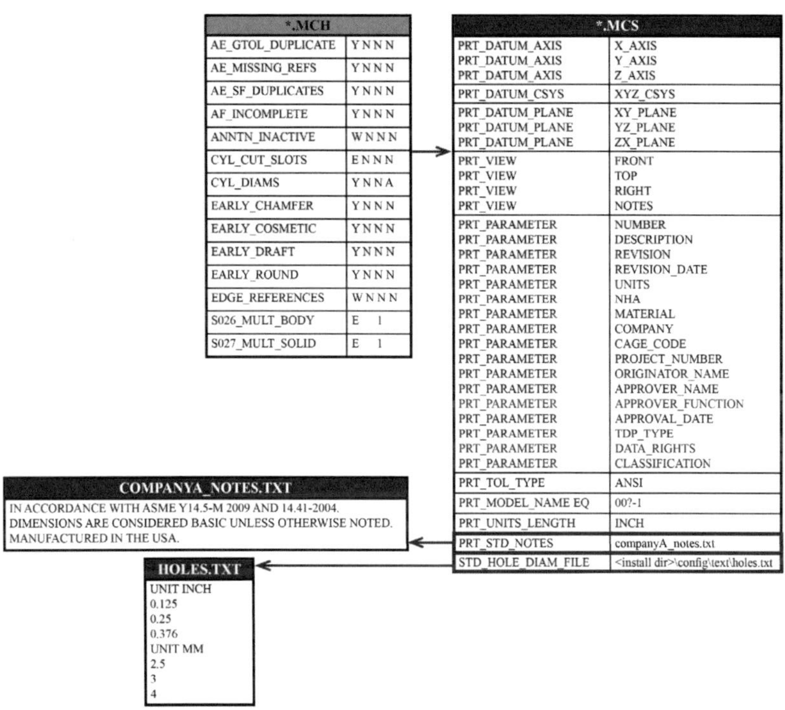


***File Mapping of ModelCHECK Files Including *.mcs and holes.txt***

# .MCN

The .mcn file, known as the constant file, holds a list of options that define the threshold of other checks run by ModelCHECK. The format is simple: an option followed by a value, which is usually numeric.

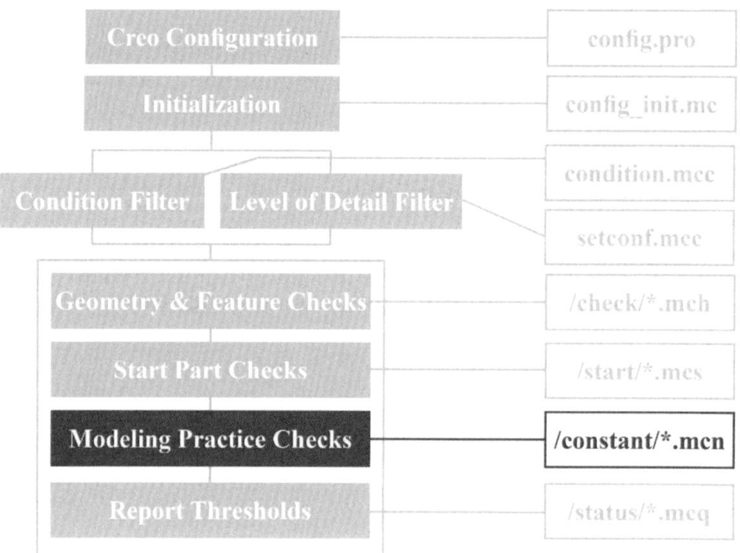


***Visual Representation of ModelCHECK Files***
***Focusing on the Modeling Practice Checks File***

When we reviewed the **.mch* file, a few options starting on page 73 looked for features like early chamfers, cosmetic features, drafts and rounds in the model tree. The exact options that I am referring to were called:

EARLY_CHAMFER
EARLY_COSMETIC
EARLY_DRAFT
EARLY_ROUND

## EARLY_CHAMFER <YNEW> Y N N N

The **.mcn* file holds the values that determine what makes these checks "true" or evaluate such that a flag is thrown.

The options that relate directly to these checks are:

PERC_EARLY_CHAMF
PERC_EARLY_COSMETIC
PERC_EARLY_DRAFT
PERC_EARLY_ROUND

If you refer back to the explanation of the options that trigger this check, you might recall that, generally, the stability of a model is better when finishing features such as chamfers are added to the model toward the end of the design process, as discussed on page 110 in the print version of Re-Use Your CAD: The Model-Based CAD Handbook, 1st Edition. So as you might have come to expect by now, I will let you determine the values to which these options should be set.

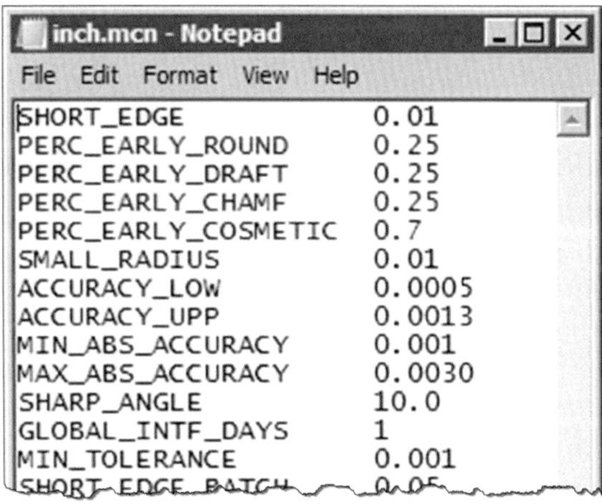


***Example *.mcn File in Notepad***

Here's a close-up of where the *.mcn file fits into the file mapping diagram.

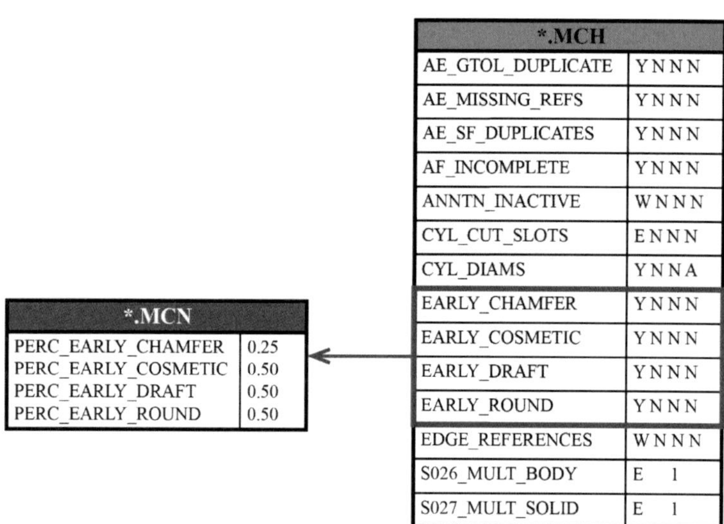


***File Mapping of ModelCHECK Files Including *.mcn***

# *.MCQ

Finally, we arrive at the last file in the ModelCHECK hierarchy, which is appropriate since the **.mcq* file is the file that determines the status of the model when the report is run, based on the results of the checks that have been set.

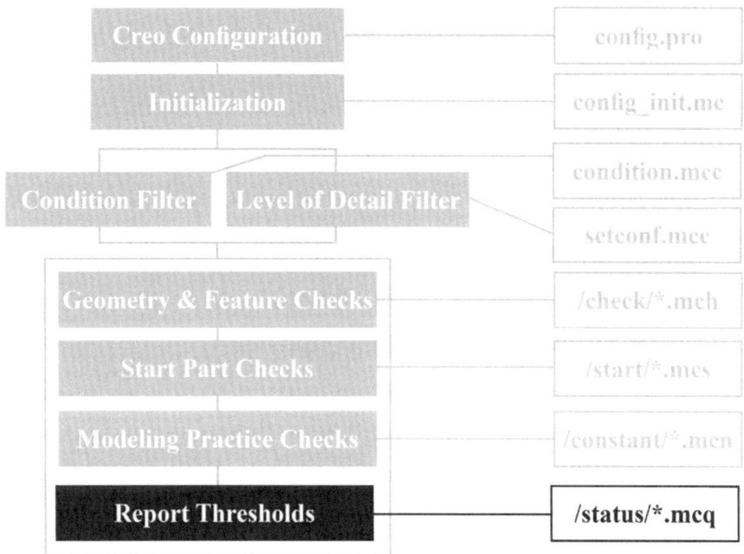


***Visual Representation of ModelCHECK Files***
***Focusing on the Report Threshold File***

You could also think of this check as automated "CAD COP" because it sets the limits for acceptable, warning and failure (or not allowed). Three colors are used as indicators for model status: green, yellow and red. You will decide how many errors and warnings are allowed, which will then determine the overall status of the model. Let's take a look at the file:

sample_status.mcq - Notepad

File Edit Format View Help

```
GREEN    5E,10W
YELLOW   8E,15W
```

***Example *.mcq File in Notepad***

This file will:

- Set the status of the model to green if there are less than 5 errors and/or 10 warnings.
- Set the status of the model to yellow if there are between 5 – 8 errors and/or 10 - 15 warnings.
- Set the status of the model to **red** if there are more than 8 errors and/or more than 15 warnings.

Because the pass / fail criteria are highly dependent on the configuration level you set up and run in ModelCHECK, every change made to the configuration should be evaluated against how the red / yellow / green reporting will be affected.

And here's the final piece to the ModelCHECK file mapping diagram.

| *.MCS | |
|---|---|
| PRT_DATUM_AXIS | X_AXIS |
| PRT_DATUM_AXIS | Y_AXIS |
| PRT_DATUM_AXIS | Z_AXIS |
| PRT_DATUM_CSYS | XYZ_CSYS |
| PRT_DATUM_PLANE | XY_PLANE |
| PRT_DATUM_PLANE | YZ_PLANE |
| PRT_DATUM_PLANE | ZX_PLANE |
| PRT_VIEW | FRONT |
| PRT_VIEW | TOP |
| PRT_VIEW | RIGHT |
| PRT_VIEW | NOTES |
| PRT_PARAMETER | NUMBER |
| PRT_PARAMETER | DESCRIPTION |
| PRT_PARAMETER | REVISION |
| PRT_PARAMETER | REVISION_DATE |
| PRT_PARAMETER | UNITS |
| PRT_PARAMETER | NHA |
| PRT_PARAMETER | MATERIAL |
| PRT_PARAMETER | COMPANY |
| PRT_PARAMETER | CAGE_CODE |
| PRT_PARAMETER | PROJECT_NUMBER |
| PRT_PARAMETER | ORIGINATOR_NAME |
| PRT_PARAMETER | APPROVER_NAME |
| PRT_PARAMETER | APPROVER_FUNCTION |
| PRT_PARAMETER | APPROVAL_DATE |
| PRT_PARAMETER | TDP_TYPE |
| PRT_PARAMETER | DATA_RIGHTS |
| PRT_PARAMETER | CLASSIFICATION |
| PRT_TOL_TYPE | ANSI |
| PRT_MODEL_NAME EQ | 00?-1 |
| PRT_UNITS_LENGTH | INCH |
| PRT_STD_NOTES | companyA_notes.txt |
| STD_HOLE_DIAM_FILE | <install dir>\config\text\holes.txt |

| *.MCQ | |
|---|---|
| GREEN | 5E, 10W |
| YELLOW | 8E, 15W |

***File Mapping of ModelCHECK Files Including *.mcq***

On the next two pages I've provided a larger image in an effort to display all of the file relations in one place.

A larger version of this image in PDF, JPG and PNG formats can be found on the Action Engineering website at www.action-engineer.com/store.

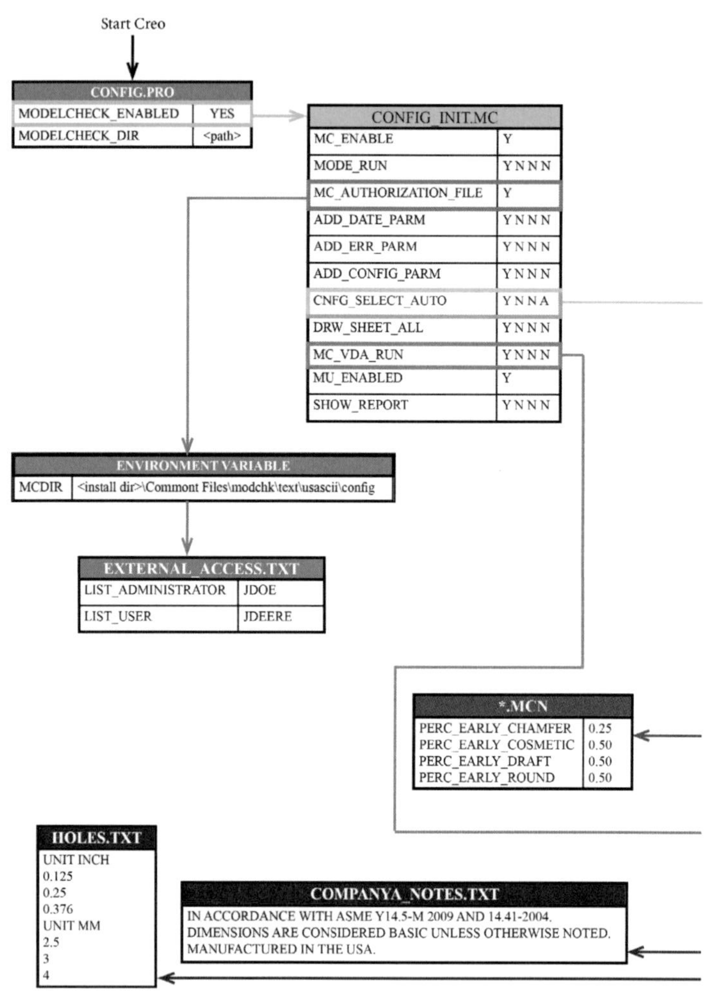
Start Creo
CONFIG.PRO
MODELCHECK_ENABLED YES
MODELCHECK_DIR <path>
CONFIG_INIT.MC
MC_ENABLE Y
MODE_RUN Y N N N
MC_AUTHORIZATION_FILE Y
ADD_DATE_PARM Y N N N
ADD_ERR_PARM Y N N N
ADD_CONFIG_PARM Y N N N
CNFG_SELECT_AUTO Y N N A
DRW_SHEET_ALL Y N N N
MC_VDA_RUN Y N N N
MU_ENABLED Y
SHOW_REPORT Y N N N
ENVIRONMENT VARIABLE
MCDIR <install dir>\Commont Files\modchk\text\usascii\config
EXTERNAL_ACCESS.TXT
LIST_ADMINISTRATOR JDOE
LIST_USER JDEERE
*.MCN
PERC_EARLY_CHAMFER 0.25
PERC_EARLY_COSMETIC 0.50
PERC_EARLY_DRAFT 0.50
PERC_EARLY_ROUND 0.50
HOLES.TXT
UNIT INCH
0.125
0.25
0.376
UNIT MM
2.5
3
4
COMPANYA_NOTES.TXT
IN ACCORDANCE WITH ASME Y14.5-M 2009 AND 14.41-2004.
DIMENSIONS ARE CONSIDERED BASIC UNLESS OTHERWISE NOTED.
MANUFACTURED IN THE USA.

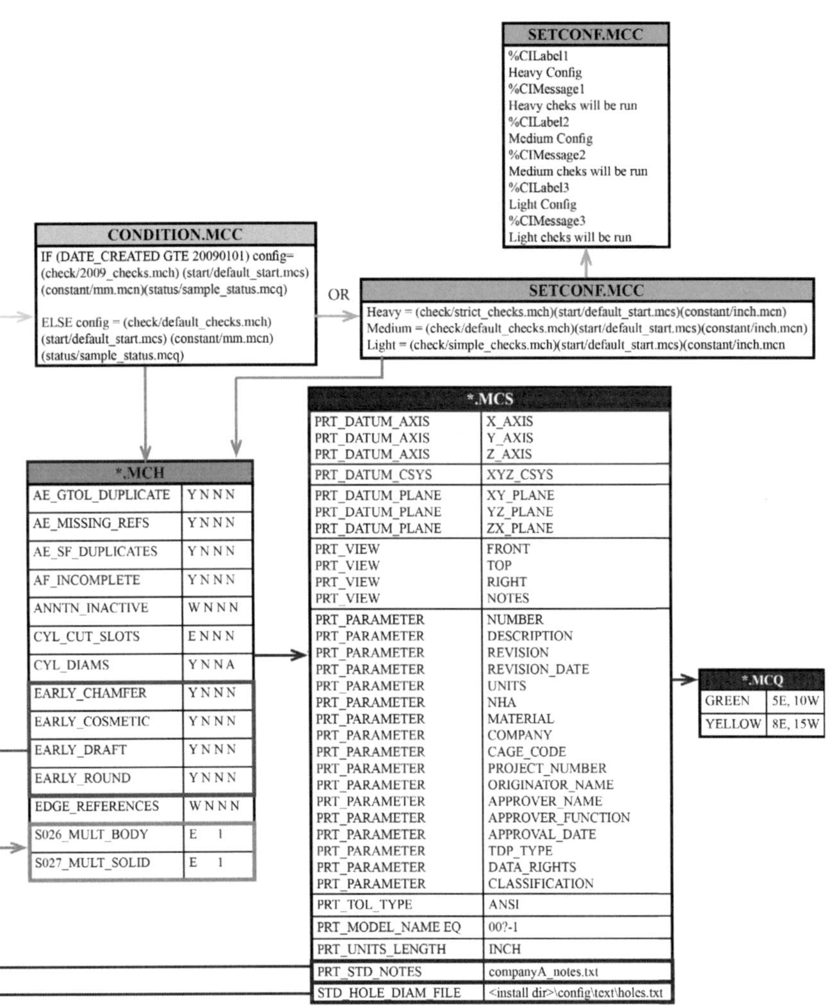
SETCONF.MCC
%CILabel1
Heavy Config
%CIMessage1
Heavy cheks will be run
%CILabel2
Medium Config
%CIMessage2
Medium cheks will be run
%CILabel3
Light Config
%CIMessage3
Light cheks will be run
CONDITION.MCC
IF (DATE_CREATED GTE 20090101) config=
(check/2009_checks.mch) (start/default_start.mcs)
(constant/mm.mcn)(status/sample_status.mcq)
ELSE config = (check/default_checks.mch)
(start/default_start.mcs) (constant/mm.mcn)
(status/sample_status.mcq)
OR
SETCONF.MCC
Heavy = (check/strict_checks.mch)(start/default_start.mcs)(constant/inch.mcn)
Medium = (check/default_checks.mch)(start/default_start.mcs)(constant/inch.mcn)
Light = (check/simple_checks.mch)(start/default_start.mcs)(constant/inch.mcn
*.MCH
AE_GTOL_DUPLICATE | Y N N N
AE_MISSING_REFS | Y N N N
AE_SF_DUPLICATES | Y N N N
AF_INCOMPLETE | Y N N N
ANNTN_INACTIVE | W N N N
CYL_CUT_SLOTS | E N N N
CYL_DIAMS | Y N N A
EARLY_CHAMFER | Y N N N
EARLY_COSMETIC | Y N N N
EARLY_DRAFT | Y N N N
EARLY_ROUND | Y N N N
EDGE_REFERENCES | W N N N
S026_MULT_BODY | E 1
S027_MULT_SOLID | E 1
*.MCS
PRT_DATUM_AXIS | X_AXIS
PRT_DATUM_AXIS | Y_AXIS
PRT_DATUM_AXIS | Z_AXIS
PRT_DATUM_CSYS | XYZ_CSYS
PRT_DATUM_PLANE | XY_PLANE
PRT_DATUM_PLANE | YZ_PLANE
PRT_DATUM_PLANE | ZX_PLANE
PRT_VIEW | FRONT
PRT_VIEW | TOP
PRT_VIEW | RIGHT
PRT_VIEW | NOTES
PRT_PARAMETER | NUMBER
PRT_PARAMETER | DESCRIPTION
PRT_PARAMETER | REVISION
PRT_PARAMETER | REVISION_DATE
PRT_PARAMETER | UNITS
PRT_PARAMETER | NHA
PRT_PARAMETER | MATERIAL
PRT_PARAMETER | COMPANY
PRT_PARAMETER | CAGE_CODE
PRT_PARAMETER | PROJECT_NUMBER
PRT_PARAMETER | ORIGINATOR_NAME
PRT_PARAMETER | APPROVER_NAME
PRT_PARAMETER | APPROVER_FUNCTION
PRT_PARAMETER | APPROVAL_DATE
PRT_PARAMETER | TDP_TYPE
PRT_PARAMETER | DATA_RIGHTS
PRT_PARAMETER | CLASSIFICATION
PRT_TOL_TYPE | ANSI
PRT_MODEL_NAME EQ | 00?-1
PRT_UNITS_LENGTH | INCH
PRT_STD_NOTES | companyA_notes.txt
STD_HOLE_DIAM_FILE | <install dir>\config\text\holes.txt
*.MCQ
GREEN | 5E, 10W
YELLOW | 8E, 15W

# PART FOUR: READING THE MODELCHECK REPORT

Now that we've covered all of the details of the files and the checks that can be set in the files, we need to understand how to interpret the results that ModelCHECK will report back in the browser embedded in Creo (see image below). Results are blurred out so you don't get hung up on the details of what is being reported; we will get to that! If you recall, back in the *config_init.mc* file, an option named SHOW_REPORT, set to Y by default, must be turned on in order for the browser to automatically show the results. If this is set to N, the HTML content of the report is still generated, but you will have to locate it and open it manually to view the results.

Notice that you can still view the model in the graphics window on the right and the model tree on the left, which will be helpful when we sift through the detailed results.

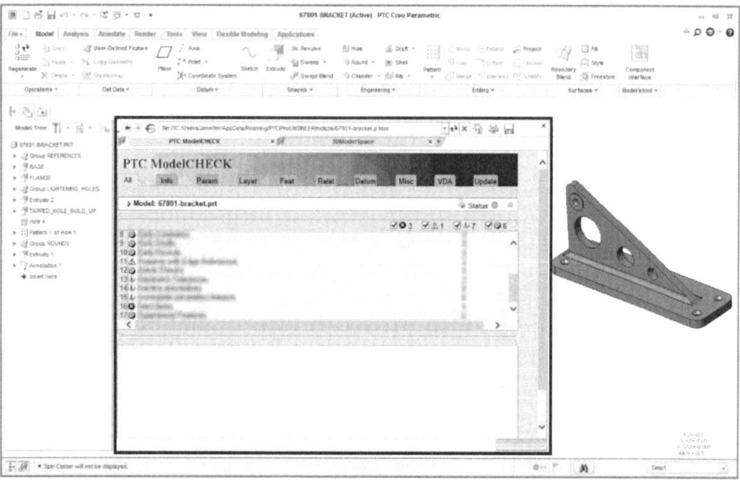


***Example ModelCHECK Report***

Starting at the top of the embedded browser, the report will show several tabs that list the categories of the types of checks that ModelCHECK runs. By default, the 'All' tab is activated, which means it will show checks of all types in the bottom half of the browser window, but these tabs can be useful to quickly filter out the type of information that you will be viewing in case you are looking for something specific. I've included a list of what the tab labels represent, since most of them are abbreviated.

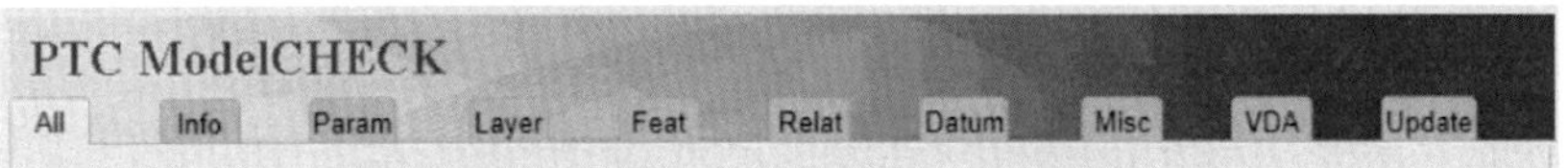


***Tabs Displayed in the ModelCHECK Report***

| Tab Label | Description | Explanation |
|---|---|---|
| All | All Checks | Shows all checks |
| Info | Information Checks | Shows the results of checks that are considered to be informative. |
| Param | Parameter Checks | Shows the results of checks performed on parameters. |
| Layer | Layer Checks | Shows the results of checks performed on layers. |
| Feat | Feature Checks | Shows the results of checks performed on features. |
| Relat | Relation Checks | Shows the results of checks performed on relations. |
| Datum | Datum Checks | Shows the results of checks performed on datum features. |
| Misc | Miscellaneous Checks | All other checks that don't fit into one of the above categories |
| VDA* | Verband der Automobilindustrie 4955 Design Specification Checks | Checks geometry for conformance to the VDA standard. The same checks are run when GeomIntegrityCHECK is run. |
| Update | Model Update | All checks that update the model but do not require user interaction. |

***Explanation of the Tab Labels in the ModelCHECK Report***

In 2001, a bug was reported that indicated results are not being written to the VDA tab in the ModelCHECK report. Details can be found on the PTC support website and are documented under Technical Application Note (TAN) 123736.

The next line of information shown contains the name of the file that ModelCHECK was run against (on the left) and the overall status of that file (on the right). If running ModelCHECK in Interactive mode, you would know the file that it was run against. But if you use another mode, such as Batch or Save, you may choose to review the results at a later time, in which point this information becomes more relevant.

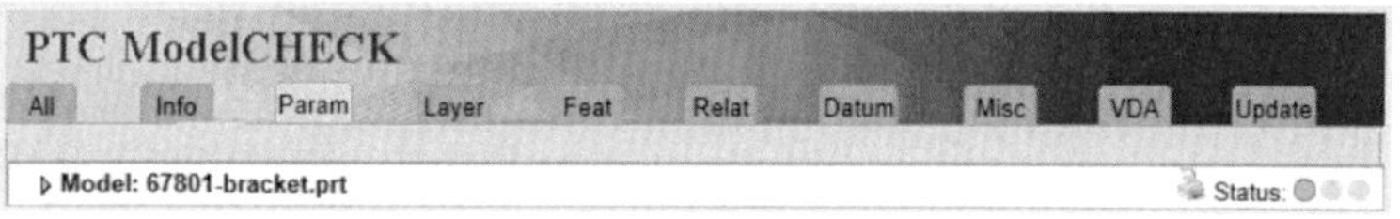


***Name of the File that ModelCHECK was Run Against***

The overall status of the file is determined by the values that were set in the **.mcn* file. This is where you set the threshold for the number of errors and warnings, which determines the status of the model overall. As you might expect, green is a sign that everything checks out okay, yellow indicates that some checks may require investigation and red is a sure sign that something is not right. The table below describes the actual conditions evaluated to determine model status.

| Icon | Meaning |
|---|---|
| ● ○ ○ | The total number of errors and warnings in the checked file is less than the number specified to indicate a GREEN status in the **.mcq* file. |
| ○ ● ○ | The total number of errors and warnings in the checked file is more than the number specified for a GREEN status, but less than the number specified to indicate a RED status in the **.mcq* file. |
| ○ ○ ● | The total number of errors and warnings in the checked file is more than the number specified to indicate a YELLOW status in the **.mcq* file. |

***Explanation of Status Icons in the ModelCHECK Report***

This is a quick way to view the overall status of the file, but how do you know which checks were run? Click the arrow next to the word **Model**, and the window will expand to show the details of when and how ModelCHECK was run, who ran it and which ModelCHECK files were read. The following image shows this.

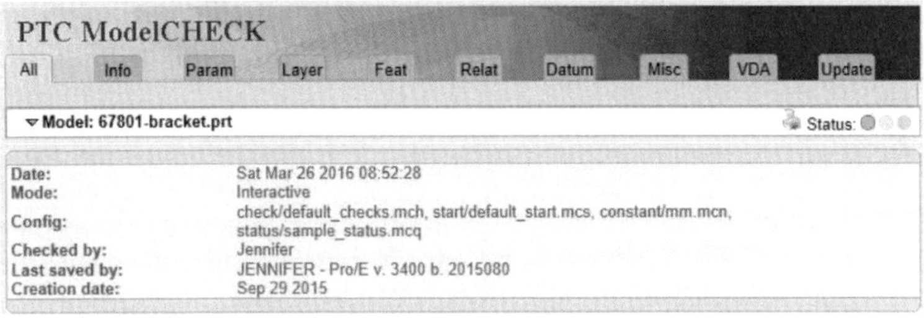

*Details of When and How ModelCHECK was Run*

The next pane shows a list of the checks that were run, the status of each check (to the left of the check name) and the total number of that check that was evaluated (under the Result heading). In the top right corner of this pane, you will see numbers that represent the sum total of how many checks failed, how many resulted in a warning, how many were informational and how many passed, respectively.

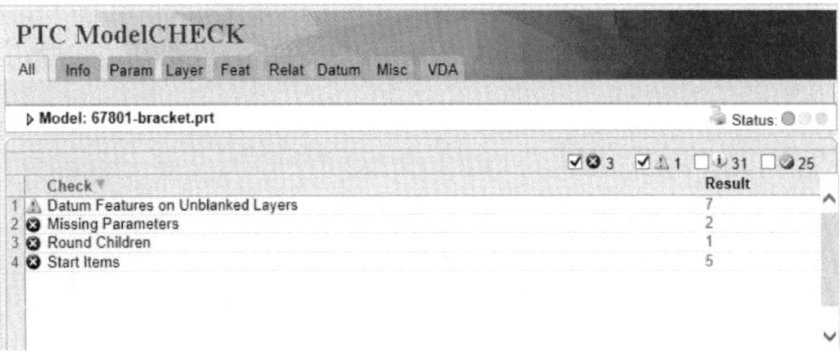

*List of Checks that Were Run and Their Status*

By default, this pane of the report will open up on the "All" tab, and will only show checks that resulted in an error or warning, as shown in the previous image. You can customize which checks are shown by checking the box(es) next to the icon of the status(es) that you want to display. The next image shows the same report, but with all status icons enabled.

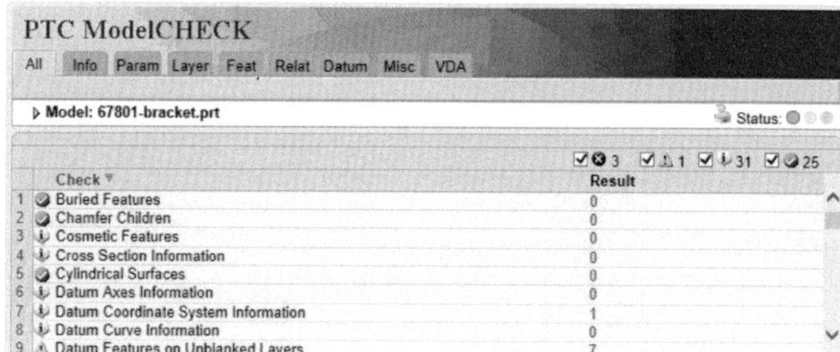


***ModelCHECK Report with Checks of all Statuses Displayed***

What determines if a check passes, fails, is informational or a warning? Remember the "YNEW" settings in the various configuration files? They finally get explained! The following table explains how pass, fail, informational and warning statuses are determined.

| Icon | Description | Explanation |
|---|---|---|
| | Pass | The check was set to Y in the appropriate configuration file and met the specified criteria. |
| | Information | Information about the check that was run is shown. |
| | Warning | The check was set to W in the appropriate configuration file and did not meet the specified criteria. |
| | Failure | The check was set to E in the appropriate configuration file and did not meet the specified criteria. |

***Explanation of How a Check Status is Determined***

To view detailed information for each check, simply select that check for details to be displayed. This final panel in the report window is where you can interact with the report to highlight problem areas in the model or add or change information, such as parameters. Let's take a look at a few examples to help you visualize this. The information presented for each check will vary slightly, but once you have the gist of what is being shown, it is relatively intuitive to interpret.

First we'll take a look at a list of checks that have passed. The first check, Buried Features, reports a zero in the Result column, meaning that zero buried features were found in the model. The bottom frame provides more details. In this case: "Number of buried features: 0"

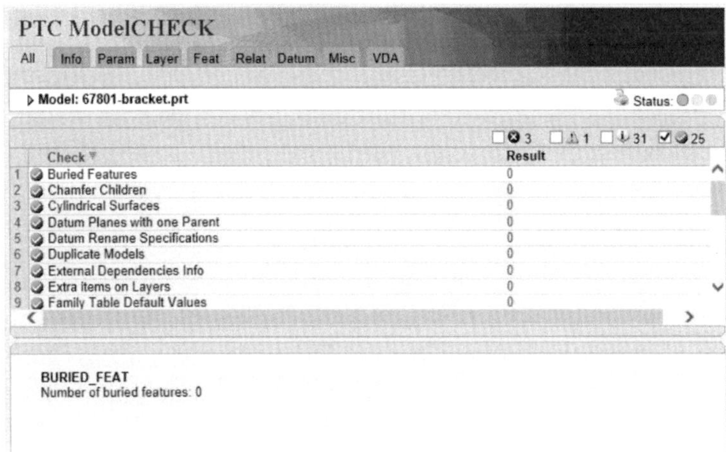


***ModelCHECK Report Showing Checks that have Passed***

The next image shows information only check. I chose to highlight the "Datum Plane Information" check, displaying a value of 6 in the Result column. The bottom frame shows that ModelCHECK found six datum planes in the model, named A, B, C, FRONT, RIGHT and TOP.

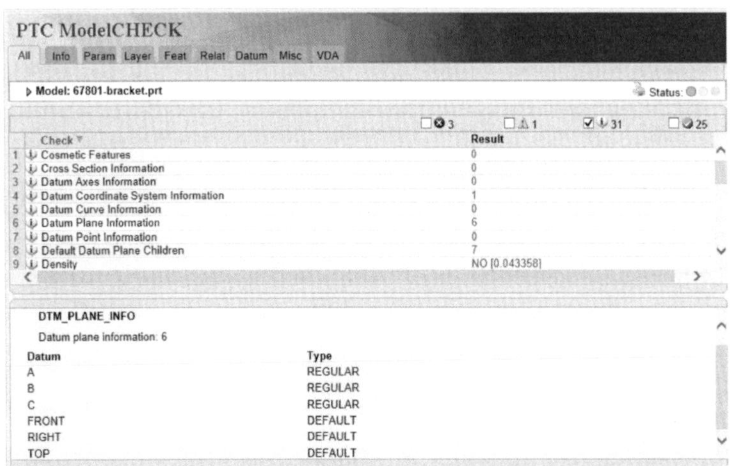


***ModelCHECK Report Showing Checks with Information Only***

You can begin interacting with the report when viewing warnings. The image below shows a warning about datum features on layers that have not been set to blank. If you highlight this row in the report, the bottom pane shows further details.

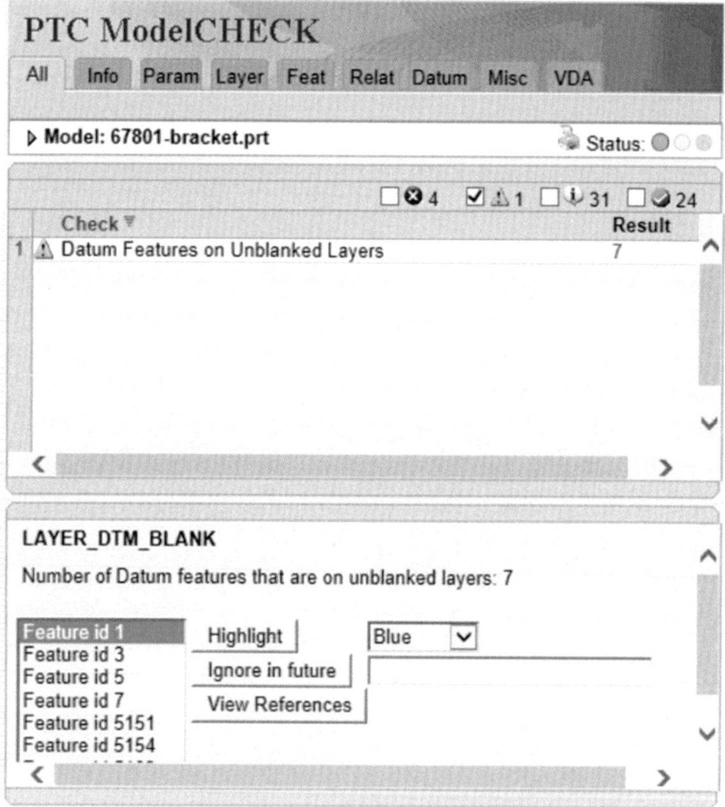


***Interaction with the ModelCHECK Report Panel***

This detailed report shows that "Feature id 1" is a datum feature on an un-blanked layer. To determine the feature to which this refers, go to the model tree to make sure that the feature ID column is displayed. In this case, the first few features are easy to identify because they are early in the feature creation list, but finding "Feature id 5151" would be more time-consuming. This is where the interactive functions in the report are helpful.

The following paragraphs describe how to interact with this result.

**Highlight:** This allows you to highlight the selected feature in a particular color. The images below show two different datum features (which both happen to be datum planes), highlighted in blue and red.

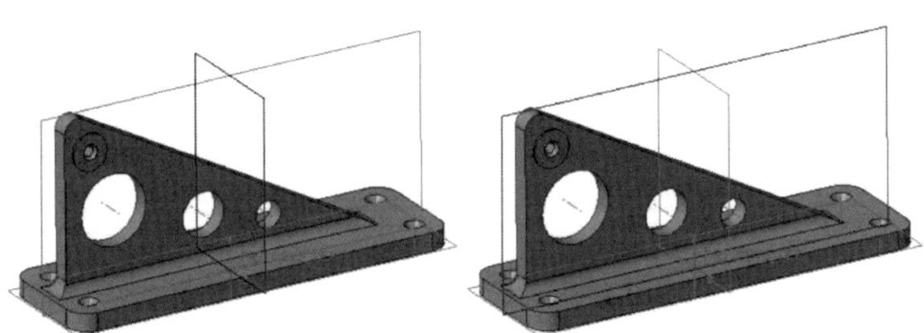

***Datum Features Highlighted in Blue and Red Using the ModelCHECK Report Panel***

**Ignore in Future:** This option allows you to turn the notification for this feature on or off when ModelCHECK runs in the future. This can be useful when you've come across an exception to a rule and don't want to set up a custom set of conditions and checks for this situation.

**View References:** This option will display the reference viewer so that you can see other features connected to the selected feature.

You also have the capability to manipulate model information directly in the report window. In the next example, a check was run to look for the existence of part parameters in the model. When ModelCHECK reports that they don't exist, you have the opportunity to add them to the model right from the report panel.

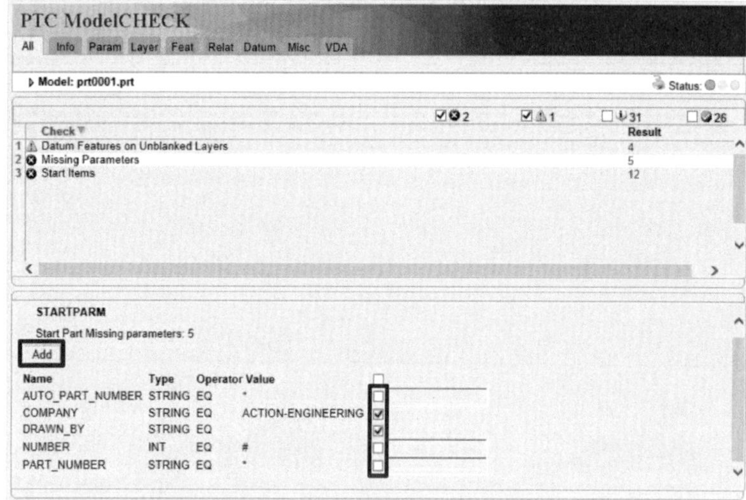


***ModelCHECK Report Adding Parameters with Values from the Report***

In the previous image, five part parameters named AUTO_PART_NUMBER, COMPANY, DRAWN_BY, NUMBER and PART_NUMBER were not present in the checked model. Place a check mark in the box to the right of the parameter information, and then press the "Add" button, the parameters will be created. Furthermore, if there is a text box to the right of the check box, you can enter a value into that field and, when the parameter is created, it will be set to that value.

# PART FIVE: DESIGNING AND DEPLOYING A MODELCHECK IMPLEMENTATION

This section is intended to provide you with an organized plan and set of guidelines to get you up and running as quickly as possible, given some pre-defined goals. The steps are arranged in a checklist format. Printable high-resolution checklists are available for download at www.action-engineering.com/store. Use these pages as a tool to check off your implementation progress.

Before you begin any implementation plan, whether it be one outlined in this book or a customized plan that you've come up with on your own, it is a good idea to make a backup of the default files that ModelCHECK uses. Yes, you can always run the PTC installer again and reinstall ModelCHECK to retrieve the default installation, but I find it far easier and faster to copy and paste the root folder of files from an existing installation. The directory named "modchk" can be found under the "Common Files" directory in the directory where Creo was installed. This is where all of the files relevant to ModelCHECK are located. For example, for an installation of Creo 3.0 M040, the "modchk" directory would be found in:

*C:\Program Files\PTC\Creo 3.0\M040\Common Files*

I'm suggesting that you copy the entire "modchk" folder and name it something like "modchk_original" or "modchk_default." Or for even more protection, move this back-up copy to another location altogether and restrict its access to administrators only. This will help to ensure that all client machines are using the same ModelCHECK configuration in order to maintain consistency in the organization.

This book was written using PTC Creo 3.0. However, you will find that the available options haven't changed in some time. Yes, there have been a few checks added here and there and, where appropriate, I will note them. Otherwise you should be able to use the majority of the checks outlined in this book across several versions of PTC Creo and Pro/ENGINEER.

# Re-Use Your CAD Implementation Checklist

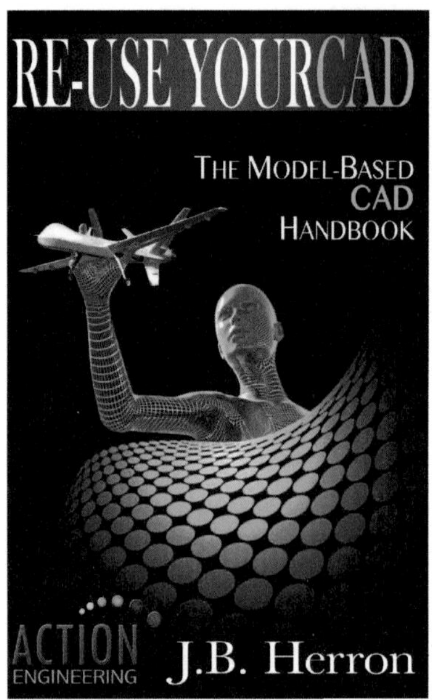


This checklist guides your ModelCHECK setup to be in compliance with Re-Use Your CAD: The Model-Based CAD Handbook, 1st Edition.

If you've read and begun to instantiate the processes presented in Re-Use Your CAD: The Model-Based CAD Handbook, 1st Edition, then you are well on your way to a model-based approach and are familiar with the rules and best practices starting on pages 82 & 94, respectively of the print version. The following pages are intended to check for as many of those recommended best practices as possible by using checks that exist in ModelCHECK today. Of course today, gaps exist between what *should* be checked and what *can* be checked with ModelCHECK. Remember ModelCHECK was around long before MBD (PTC acquired ModelCHECK from RAND Worldwide in 1999), so by using it to check a model for best MBD practices over 15 years later, we are pushing its intended boundaries a bit. However, you have it, it is free, so use it wisely.

The following checklist pages contain a list of the files you will need to edit and the changes you will need to make to them, in order to check recommended best practices for MBD. Any areas that are shown in gray are intended for you to fill in in order to customize your installation as you see fit for your organization.

# Re-Use Your CAD Implementation Checklist

<table>
<tr><th colspan="4">Preparation Checklist</th></tr>
<tr><td rowspan="2">☐</td><td colspan="3">1. List one or more goals that you want to achieve by using ModelCHECK (discussed on page 14)</td></tr>
<tr><td colspan="3"></td></tr>
<tr><td rowspan="11">☐</td><td colspan="3">2. Identify the people who need to be involved in the implementation, and their permissions (discussed on page 16)</td></tr>
<tr><td>Username</td><td>Role</td><td>Permissions Set?</td></tr>
<tr><td></td><td></td><td></td></tr>
<tr><td></td><td></td><td></td></tr>
<tr><td></td><td></td><td></td></tr>
<tr><td></td><td></td><td></td></tr>
<tr><td></td><td></td><td></td></tr>
<tr><td></td><td></td><td></td></tr>
<tr><td></td><td></td><td></td></tr>
<tr><td></td><td></td><td></td></tr>
<tr><td></td><td></td><td></td></tr>
<tr><td>☐</td><td colspan="3">3. Set permissions to allow the people identified in step 2 to make modifications to the files, or refer to the external_access.txt file on the next page to list users in an external file.</td></tr>
</table>

# Re-Use Your CAD Implementation Checklist

| config_init.mc | | |
|---|---|---|
| Make edits to the following options with a text editor or in the User Interface. | | |
| | **OPTION** | **VALUE** |
| ☐ | ADD_DATE_PARM | Y N N N |
| ☐ | ADD_ERR_PARM | Y N N N |
| ☐ | ADD_CONFIG_PARM | Y N N N |
| ☐ | CNFG_SELECT_AUTO | Y N N N |
| ☐ | DRW_SHEET_ALL | Y N N N |
| ☐ | MC_AUTHORIZATION_FILE | Y |
| ☐ | MC_ENABLE | Y |
| ☐ | MC_VDA_RUN | Y |
| ☐ | MU_ENABLED | Y |
| ☐ | SHOW_REPORT | Y N N N |

| external_access.txt | |
|---|---|
| The contents of this file will only be read when the option MC_AUTHROIZATION_FILE in the *config_init.mc* file is set to Y. Make edits to the following options with a text editor in order to grant file access to users. | |
| | **OPTION** |
| ☐ | LIST_ADMINISTRATOR |
| ☐ | <username> |
| ☐ | LIST_USER |
| ☐ | <username> |

# Re-Use Your CAD Implementation Checklist

| Environment Variable | | |
|---|---|---|
| Make edits to the following system Environment Variable. This variable will only be read when the option MC_AUTHROIZATION_FILE in the *config_init.mc* file is set to Y. | | |
| | **VARIABLE** | **VALUE** |
| ☐ | MC_DIR | <path to the folder where the external_access.txt file is located> |

| condition.mcc | | |
|---|---|---|
| Make edits to the following options with a text editor or in the User Interface. The contents of this file will only be read when the option CNFG_SELECT_AUTO in the *config_init.mc* file is set to Y. | | |
| | **OPTION** | **VALUE** |
| ☐ | *IF* ______<br>______<br>______<br>config = | (check/______.mch)<br>(start/______.mcs)<br>(constant/______.mcn)<br>(status/______.mcq) |
| ☐ | *IF* ______<br>______<br>______<br>config = | (check/______.mch)<br>(start/______.mcs)<br>(constant/______.mcn)<br>(status/______.mcq) |
| ☐ | *ELSE* ______<br>______<br>______<br>config = | (check/______.mch)<br>(start/______.mcs)<br>(constant/______.mcn)<br>(status/______.mcq) |
| ☐ | *IF (NOCHECK)* | *NOCHECK* |

# Re-Use Your CAD Implementation Checklist

| setnconf.mcc | | |
|---|---|---|
| Make edits to the following options with a text editor or in the User Interface. The contents of this file will only be read when the option CNFG_SELECT_AUTO in the *config_init.mc file* is set to N or A. Any number of configurations can be set in addition to or in place of the three default options. Use the blank lines to identify additional options that you would like to set. | | |
| | **OPTION** | **VALUE** |
| ☐ | Heavy | (check/strict_checks.mch)<br>(start/default_start.mcs)<br>(constant/inch.mcn) |
| ☐ | Medium | (check/default_checks.mch)<br>(start/default_start.mcs)<br>(constant/inch.mcn) |
| ☐ | Light | (check/simple_checks.mch)<br>(start/default_start.mcs)<br>(constant/inch.mcn) |
| ☐ | ________ | (check/________________.mch)<br>(start/________________.mcs)<br>(constant/________________.mcn) |
| ☐ | ________ | (check/________________.mch)<br>(start/________________.mcs)<br>(constant/________________.mcn) |
| ☐ | ________ | (check/________________.mch)<br>(start/________________.mcs)<br>(constant/________________.mcn) |
| ☐ | ________ | (check/________________.mch)<br>(start/________________.mcs)<br>(constant/________________.mcn) |

# Re-Use Your CAD Implementation Checklist

<table>
<tr><th colspan="3">msg_mc.txt</th></tr>
<tr><td colspan="3">Make edits to the following options with a text editor. The contents of this file will only be read when the option CNFG_SELECT_AUTO in the config_init.mc file is set to N or A. Use the blank lines to identify additional options that you would like to set.</td></tr>
<tr><td></td><td>OPTION</td><td>VALUE</td></tr>
<tr><td>☐</td><td>%CILabel1</td><td>Heavy Config</td></tr>
<tr><td>☐</td><td>%CIMessage1</td><td>Heavy checks will be run</td></tr>
<tr><td>☐</td><td>%CILabel2</td><td>Medium Config</td></tr>
<tr><td>☐</td><td>%CIMessage2</td><td>Medium checks will be run</td></tr>
<tr><td>☐</td><td>%CILabel3</td><td>Light Config</td></tr>
<tr><td>☐</td><td>%CIMessage3</td><td>Light checks will be run</td></tr>
<tr><td>☐</td><td>%CILabel4</td><td><Label for fourth option in UI></td></tr>
<tr><td>☐</td><td>%CIMessage4</td><td><Tooltip for fourth option in UI></td></tr>
<tr><td>☐</td><td>%CILabel5</td><td><Label for fifth option in UI></td></tr>
<tr><td>☐</td><td>%CIMessage5</td><td><Tooltip for fifth option in UI></td></tr>
<tr><td>☐</td><td>%CILabel6</td><td><Label for sixth option in UI></td></tr>
<tr><td>☐</td><td>%CIMessage6</td><td><Tooltip for sixth option in UI></td></tr>
<tr><td>☐</td><td>%CILabel7</td><td><Label for seventh option in UI></td></tr>
<tr><td>☐</td><td>%CIMessage7</td><td><Tooltip for seventh option in UI></td></tr>
</table>

# Re-Use Your CAD Implementation Checklist

| .mch | | |
|---|---|---|
| Make edits to the following options with a text editor or in the User Interface. Use the blank lines to identify additional options that you would like to set. | | |
| | **OPTION** | **VALUE** |
| ☐ | AE_GTOL_DUPLICATE | Y N N N |
| ☐ | AE_MISSING_REFS | Y N N N |
| ☐ | AE_SF_DUPLICATE | Y N N N |
| ☐ | AF_INCOMPLETE | Y N N N |
| ☐ | ANNTN_INACTIVE | Y N N N |
| ☐ | CYL_CUT_SLOTS | Y N N N |
| ☐ | CYL_DIAMS | Y N N N |
| ☐ | EARLY_CHAMFER | Y N N N |
| ☐ | EARLY_COSMETIC | Y N N N |
| ☐ | EARLY_DRAFT | Y N N N |
| ☐ | EARLY_ROUND | Y N N N |
| ☐ | EDGE_REFERENCES | Y N N N |
| ☐ | S026_MULT_BODY | E |
| ☐ | S027_MULT_SOLID | E |
| ☐ | Add your own options here… | |
| ☐ | | |
| ☐ | | |
| ☐ | | |
| ☐ | | |

# Re-Use Your CAD Implementation Checklist

### holes.txt

Make edits to the following options with a text editor. The contents of this file will only be read when the option CYL_DIAMS in the *.*mch* file is set to Y. Use the blank lines to identify additional options that you would like to set.

| | OPTION | VALUE |
|---|---|---|
| ☐ | INCH | The first line in the file should be units, INCH or MM. More than one unit can be specified in a file. |
| ☐ | MM | |
| ☐ | <value> | Each line after the units are standard hole diameters that are allowed. |
| ☐ | | |
| ☐ | | |
| ☐ | | |

### *.mcn

Make edits to the following options with a text editor. The contents of this file will only be read when the options EARLY_CHAMFER, EARSY_COSMETIC, EARLY_DRAFT and EARLY_ROUND in the *config_init.mc* file are set to Y respectively.

| | OPTION | VALUE |
|---|---|---|
| ☐ | PERC_EARLY_CHAMF | Decimal Number > 0 and < 1 |
| ☐ | PERC_EARLY_COSMETIC | Decimal Number > 0 and < 1 |
| ☐ | PERC_EARLY_DRAFT | Decimal Number > 0 and < 1 |
| ☐ | PERC_EARLY_ROUND | Decimal Number > 0 and < 1 |

# Re-Use Your CAD Implementation Checklist

| *.mcs | | |
|---|---|---|
| Make edits to the following options with a text editor or in the User Interface. Use the blank lines to identify additional options that you would like to set. | | |
| ☐ | ASM_DATUM_AXIS | X_AXIS |
| ☐ | ASM_DATUM_AXIS | Y_AXIS |
| ☐ | ASM_DATUM_AXIS | Z_AXIS |
| ☐ | PRT_DATUM_AXIS | X_AXIS |
| ☐ | PRT_DATUM_AXIS | Y_AXIS |
| ☐ | PRT_DATUM_AXIS | Z_AXIS |
| ☐ | ASM_DATUM_CSYS | XYZ_CSYS |
| ☐ | PRT_DATUM_ CSYS | XYZ_CSYS |
| ☐ | ASM_DATUM_PLANE | XY_PLANE |
| ☐ | ASM_DATUM_PLANE | YZ_PLANE |
| ☐ | ASM_DATUM_PLANE | ZX_PLANE |
| ☐ | PRT_DATUM_PLANE | XY_PLANE |
| ☐ | PRT_DATUM_PLANE | YZ_PLANE |
| ☐ | PRT_DATUM_PLANE | ZX_PLANE |
| ☐ | Add your own options here… | |
| ☐ | | |
| ☐ | | |
| ☐ | | |
| ☐ | | |
| ☐ | | |

# Re-Use Your CAD Implementation Checklist

| *.mcs (continued) | | |
|---|---|---|
| Make edits to the following options with a text editor or in the User Interface. Use the blank lines to identify additional options that you would like to set. | | |
| ☐ | ASM_VIEW | FRONT |
| ☐ | ASM_VIEW | TOP |
| ☐ | ASM_VIEW | RIGHT |
| ☐ | ASM_VIEW | NOTES |
| ☐ | PRT_VIEW | FRONT |
| ☐ | PRT_VIEW | TOP |
| ☐ | PRT_VIEW | RIGHT |
| ☐ | PRT_VIEW | NOTES |
| ☐ | Add your own options here… | |
| ☐ | | |
| ☐ | | |
| ☐ | | |
| ☐ | | |
| ☐ | | |
| ☐ | | |
| ☐ | | |
| ☐ | | |
| ☐ | | |
| ☐ | | |
| ☐ | | |

# Custom ModelCHECK Implementation Checklist

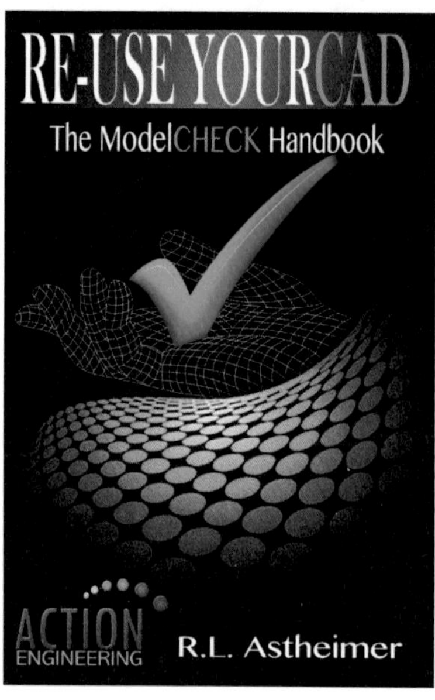


The following pages are intended to guide you through a customized ModelCHECK implementation by providing you with a list of all files involved and the options that you might want to set for an organized approach to your implementation. Any areas that are shown in gray are intended for you to fill in in order to customize your installation as you see fit for your organization.

# ModelCHECK Implementation Checklist

## For a Custom Implementation

| Preparation Checklist | |
|---|---|
| ☐ | 1. List one or more goals that you want to achieve by using ModelCHECK (discussed on page 14) |
| | |
| ☐ | 2. Identify the people who need to be involved in the implementation, and their permissions (discussed on page 16) |

| Username | Role | Permissions Set? |
|---|---|---|
| | | |
| | | |
| | | |
| | | |
| | | |
| | | |
| | | |
| | | |
| | | |

| | |
|---|---|
| ☐ | 3. Set permissions to allow the people identified in step 2 to make modifications to the files, or refer to the external_acesss.txt file on the next page to list users in an external file. |

# ModelCHECK Implementation Checklist

## For a Custom Implementation

| What do you want to check? (circle all that apply) | | |
|---|---|---|
| Parts | Assemblies | Drawings |

| What are you attempting to check for? How? (check all that apply and/or add to the list) | |
|---|---|
| ☐ | Good modeling practice |
| | *Example: The model should pass GeomIntegrity checks* |
| ☐ | |
| ☐ | |
| ☐ | |
| ☐ | Compliance to a standard |
| | *Example: Check the units that are set* |
| ☐ | |
| ☐ | |
| ☐ | |
| ☐ | Completeness of a data set |
| | *Example: All parameters exist in file* |
| ☐ | |
| ☐ | |
| ☐ | Other |
| ☐ | |
| ☐ | |

# ModelCHECK Implementation Checklist

## For a Custom Implementation

| config_init.mc | | |
|---|---|---|
| The following options are required for ModelCHECK to run. Edits can be made with a text editor or in the User Interface. Additional options are available in this file; use the blank lines to identify additional options that you would like to set. | | |
| | **OPTION** | **VALUE** |
| ☐ | CNFG_SELECT_AUTO | Y N N N |
| ☐ | MC_AUTHORIZATION_FILE | Y |
| ☐ | MC_ENABLE | Y |
| ☐ | MC_VDA_RUN | Y |
| ☐ | MU_ENABLED | Y |
| ☐ | SHOW_REPORT | Y N N N |
| ☐ | Add your own options here… | |
| ☐ | | |
| ☐ | | |
| ☐ | | |
| ☐ | | |
| ☐ | | |
| ☐ | | |
| ☐ | | |
| ☐ | | |
| ☐ | | |
| ☐ | | |
| ☐ | | |

# ModelCHECK Implementation Checklist

## For a Custom Implementation

| external_access.txt | |
|---|---|
| The contents of this file will only be read when the option MC_AUTHROIZATION_FILE in the *config_init.mc* file is set to Y. Make edits to the following options with a text editor in order to grant file access to users. | |
| | **OPTION** |
| ☐ | LIST_ADMINISTRATOR |
| ☐ | Add your administrators here… |
| ☐ | |
| ☐ | LIST_USER |
| ☐ | Add your users here… |
| ☐ | |
| ☐ | |
| ☐ | |
| ☐ | |
| ☐ | |
| ☐ | |
| ☐ | |
| ☐ | |
| ☐ | |
| ☐ | |
| ☐ | |
| ☐ | |

# ModelCHECK Implementation Checklist

## For a Custom Implementation

| Environment Variable | | |
|---|---|---|
| Make edits to the following system Environment Variable. This variable will only be read when the option MC_AUTHROIZATION_FILE in the *config_init.mc* file is set to Y. | | |
| | **VARIABLE** | **VALUE** |
| ☐ | MC_DIR | ______________________<br>______________________<br><path to the folder where the external_access.txt file is located> |

# ModelCHECK Implementation Checklist

## For a Custom Implementation

| condition.mcc | | |
|---|---|---|
| Make edits to the conditional statements, based on your needs, with a text editor or in the User Interface. The contents of this file will only be read when the option CNFG_SELECT_AUTO in the *config_init.mc* file is set to Y. Use blank lines to identify additional conditions that you would like to check. | | |
| | **OPTION** | **VALUE** |
| ☐ | *IF*__________<br>__________<br>__________<br>config = | (check/________.mch)<br>(start/________.mcs)<br>(constant/________.mcn)<br>(status/________.mcq) |
| ☐ | *IF*__________<br>__________<br>__________<br>config = | (check/________.mch)<br>(start/________.mcs)<br>(constant/________.mcn)<br>(status/________.mcq) |
| ☐ | *IF*__________<br>__________<br>__________<br>config = | (check/________.mch)<br>(start/________.mcs)<br>(constant/________.mcn)<br>(status/________.mcq) |
| ☐ | *IF*__________<br>__________<br>__________<br>config = | (check/________.mch)<br>(start/________.mcs)<br>(constant/________.mcn)<br>(status/________.mcq) |
| ☐ | *ELSE*__________<br>__________<br>__________<br>config = | (check/________.mch)<br>(start/________.mcs)<br>(constant/________.mcn)<br>(status/________.mcq) |
| ☐ | *IF (NOCHECK)* | *NOCHECK* |

# ModelCHECK Implementation Checklist

## For a Custom Implementation

| setnconf.mcc | | |
|---|---|---|
| Make edits to the following options with a text editor or in the User Interface. The contents of this file will only be read when the option CNFG_SELECT_AUTO in the *config_init.mc file* is set to N or A. Any number of configurations can be set in addition to or in place of the three default options. | | |
| | **OPTION** | **VALUE** |
| ☐ | Heavy | (check/strict_checks.mch)<br>(start/default_start.mcs)<br>(constant/inch.mcn) |
| ☐ | Medium | (check/default_checks.mch)<br>(start/default_start.mcs)<br>(constant/inch.mcn) |
| ☐ | Light | (check/simple_checks.mch)<br>(start/default_start.mcs)<br>(constant/inch.mcn) |
| ☐ | ________ | (check/________.mch)<br>(start/________.mcs)<br>(constant/________.mcn) |
| ☐ | ________ | (check/________.mch)<br>(start/________.mcs)<br>(constant/________.mcn) |
| ☐ | ________ | (check/________.mch)<br>(start/________.mcs)<br>(constant/________.mcn) |
| ☐ | ________ | (check/________.mch)<br>(start/________.mcs)<br>(constant/________.mcn) |

# ModelCHECK Implementation Checklist

## For a Custom Implementation

| msg_mc.txt | | |
|---|---|---|
| Make edits to the following options with a text editor. The contents of this file will only be read when the option CNFG_SELECT_AUTO in the *config_init.mc* file is set to N or A. | | |
| | OPTION | VALUE |
| ☐ | %CILabel1 | <Label for first option in UI> |
| ☐ | %CIMessage1 | <Tooltip for first option in UI> |
| ☐ | %CILabel2 | <Label for second option in UI> |
| ☐ | %CIMessage2 | <Tooltip for second option in UI> |
| ☐ | %CILabel3 | <Label for third option in UI> |
| ☐ | %CIMessage3 | <Tooltip for third option in UI> |
| ☐ | %CILabel4 | <Label for fourth option in UI> |
| ☐ | %CIMessage4 | <Tooltip for fourth option in UI> |
| ☐ | %CILabel5 | <Label for fifth option in UI> |
| ☐ | %CIMessage5 | <Tooltip for fifth option in UI> |

# ModelCHECK Implementation Checklist

## For a Custom Implementation

| .mch | | |
|---|---|---|
| Make edits to the following options with a text editor or in the User Interface. | | |
| | **OPTION** | **VALUE** |
| ☐ | AE_GTOL_DUPLICATE | Y N N N |
| ☐ | AE_MISSING_REFS | Y N N N |
| ☐ | AE_SF_DUPLICATE | Y N N N |
| ☐ | AF_INCOMPLETE | Y N N N |
| ☐ | ANNTN_INACTIVE | Y N N N |
| ☐ | CYL_CUT_SLOTS | Y N N N |
| ☐ | CYL_DIAMS | Y N N N |
| ☐ | EARLY_CHAMFER | Y N N N |
| ☐ | EARLY_COSMETIC | Y N N N |
| ☐ | EARLY_DRAFT | Y N N N |
| ☐ | EARLY_ROUND | Y N N N |
| ☐ | EDGE_REFERENCES | Y N N N |
| ☐ | S026_MULT_BODY | E |
| ☐ | S027_MULT_SOLID | E |
| ☐ | Add your own options here… | |
| ☐ | | |
| ☐ | | |
| ☐ | | |
| ☐ | | |

# ModelCHECK Implementation Checklist

## For a Custom Implementation

| holes.txt | | |
|---|---|---|
| Make edits to the following options with a text editor. The contents of this file will only be read when the option CYL_DIAMS in the **.mch* file is set to Y. | | |
| | **OPTION** | **VALUE** |
| ☐ | INCH | The first line in the file should be the units, INCH or MM. |
| ☐ | MM | |
| ☐ | <value> | Each line after the above unit are standard hole diameters that are allowed when those units are in use. |
| ☐ | | |
| ☐ | | |
| ☐ | | |
| ☐ | | |
| ☐ | INCH OR MM | More than one unit can be specified in a file. |
| ☐ | <value> | Each line after the above unit are standard hole diameters that are allowed when those units are in use. |
| ☐ | | |
| ☐ | | |
| ☐ | | |
| ☐ | | |
| ☐ | | |
| ☐ | | |
| ☐ | | |

# ModelCHECK Implementation Checklist

## For a Custom Implementation

| *.mcn | | |
|---|---|---|
| Make edits to the following options with a text editor. The contents of this file will only be read when the options EARLY_CHAMFER, EARSY_COSMETIC, EARLY_DRAFT and EARLY_ROUND in the *config_init.mc* file are set to Y respectively. Use blank lines to list additional options that you would like to set in the file | | |
| | **OPTION** | **VALUE** |
| ☐ | PERC_EARLY_CHAMF | Decimal Number > 0 and < 1 |
| ☐ | PERC_EARLY_COSMETIC | Decimal Number > 0 and < 1 |
| ☐ | PERC_EARLY_DRAFT | Decimal Number > 0 and < 1 |
| ☐ | PERC_EARLY_ROUND | Decimal Number > 0 and < 1 |
| ☐ | Add your own options here… | |
| ☐ | | |
| ☐ | | |
| ☐ | | |
| ☐ | | |
| ☐ | | |
| ☐ | | |
| ☐ | | |
| ☐ | | |
| ☐ | | |
| ☐ | | |

# ModelCHECK Implementation Checklist

## For a Custom Implementation

| *.mcs | | |
|---|---|---|
| Make edits to the following options with a text editor or in the User Interface. | | |
| ☐ | ASM_DATUM_AXIS | Add your own naming convention here… |
| ☐ | ASM_DATUM_AXIS | |
| ☐ | ASM_DATUM_AXIS | |
| ☐ | PRT_DATUM_AXIS | |
| ☐ | PRT_DATUM_AXIS | |
| ☐ | PRT_DATUM_AXIS | |
| ☐ | ASM_DATUM_CSYS | |
| ☐ | PRT_DATUM_ CSYS | |
| ☐ | ASM_DATUM_PLANE | |
| ☐ | ASM_DATUM_PLANE | |
| ☐ | ASM_DATUM_PLANE | |
| ☐ | PRT_DATUM_PLANE | |
| ☐ | PRT_DATUM_PLANE | |
| ☐ | PRT_DATUM_PLANE | |
| ☐ | Add your own options here… | |
| ☐ | | |
| ☐ | | |
| ☐ | | |
| ☐ | | |

# ModelCHECK Implementation Checklist

## For a Custom Implementation

| *.mcs (continued) | | |
|---|---|---|
| Make edits to the following options with a text editor or in the User Interface. | | |
| ☐ | ASM_VIEW | Add your own naming convention here… |
| ☐ | ASM_VIEW | |
| ☐ | ASM_VIEW | |
| ☐ | ASM_VIEW | |
| ☐ | PRT_VIEW | |
| ☐ | PRT_VIEW | |
| ☐ | PRT_VIEW | |
| ☐ | PRT_VIEW | |
| ☐ | Add your own options here… | |
| ☐ | | |
| ☐ | | |
| ☐ | | |
| ☐ | | |
| ☐ | | |
| ☐ | | |
| ☐ | | |
| ☐ | | |
| ☐ | | |
| ☐ | | |

# COMMONLY ASKED QUESTIONS

## How Do I Tell Which Files Are Being Read By ModelCHECK?

When a report is generated, the name of the file being checked can be "expanded" by clicking on the arrow to the left of its name. The row labeled "Config" shows the configuration, or in simpler terms, the list of files read when the check was run. Other useful information in this view includes when, how and who last ran ModelCHECK.

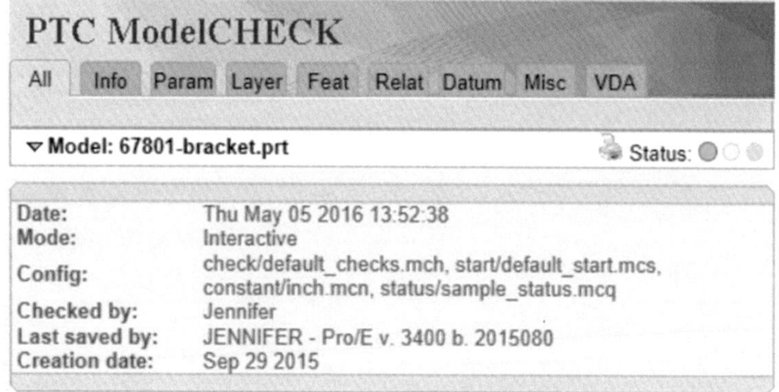


***Image Showing ModelCHECK Configuration Information in the Report***

## Does ModelCHECK Verify the Validity of Geometric Dimension & Tolerancing Information?

No. ModelCHECK does not look at the information inside a feature control frame or any associated references (such as set datums) to check for valid content.

If you would like to check your annotations for validity there is a config.pro option "restricted_gtol_dialog" that can be set to "yes". This won't check what you already have defined but it will control restrictions in the Geometric Tolerance dialog according to the selected standard when the user is defining the geometric tolerances.

## Can ModelCHECK Confirm that a Drawing Is Up-To-Date and is Synchronized with the Latest Version of the Model It is Linked To?

There is an option called MODELS_USED in the *.mch file that will list all models that are used in the drawing that ModelCHECK was run against and what sheet they are displayed on.

There is also a check called UNUSED_MODELS in the *.mch file that will list any models that are linked to the drawing file that is being checked but not shown anywhere in the drawing.

## What Does GeomIntegrityCHECK Do?

The GeomIntegrityCHECK utility checks model geometry for geometric conditions that could make transfer of the model to downstream applications difficult or require rework. The Verband der Automobilindustrie (VDA) 4955, which is an automotive standard for CAD model quality, is the standard that governs the checks that are run. There are a total of 30 checks that can be set to run, several of which their values can be customized. The image below shows the User Interface where surface checks are turned on, the types of things that are being checked for and how the value is customized.

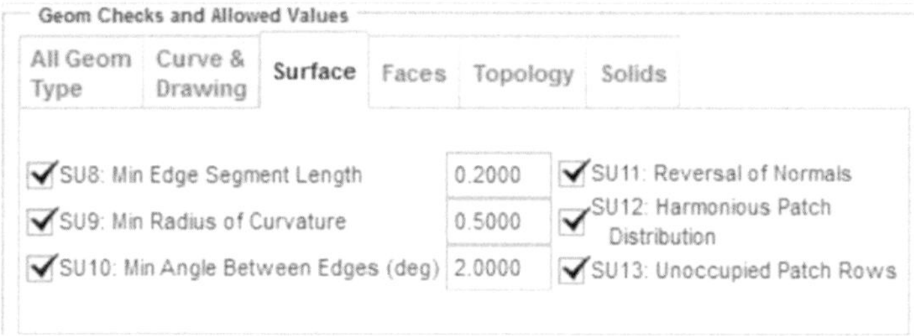


***Image of Surface Checks in the GeomIntegrityCHECK User Interface***

## Is There a Printable Version of the Implementation Plans Available For My Use?

At www.action-enginering.com/store, you will find a .pdf of the implementation plans in an 8.5 x 11 format that can be printed and used as a checklist when you go through an implementation. To download these files, enter the code **AESetupMyMC**.

## Are There Any CAD Files That I Can Use To Validate Checks That Have Been Setup in ModelCHECK?

At www.action-engineering.com/store, we provide CAD models that can be purchased and used to verify that particular ModelCHECK checks have been triggered and set up correctly. If you have any problems locating the files, or if you have a request for a specific file or check, please let us know by emailing modelcheck@action-engineering.com.

## Can I Purchase a Set of Files to be Used to Set Up an Implementation Out of the Box Based on Re-Use Your CAD: The Model-Based CAD Handbook, 1st Edition?

If you would like to purchase a set of ModelCHECK files that have been configured according to the implementation plan based on Re-Use Your CAD: The Model-Based CAD Handbook, 1st Edition, you can find them at www.action-engineering.com/store. If you have any problems locating the files, please let us know by emailing modelcheck@action-engineering.com.

# EPILOGUE

I hope you found this book helpful in unraveling the mystery of the many files involved with a basic ModelCHECK implementation. As I mentioned early on, there are a lot of files, and you might not get everything the first time, but you made it through the book! Please email any comments, suggested changes or stories from the trenches that you'd like to share about your experience and this book to modelcheck@action-engineering.com.

If you're successfully up and running after reading this book (good for you!). If you are ready to move to the next level of implementation, I'd recommend taking a look at ModelCHECK Extensions for PTC Creo offered by Software Factory, GmbH (www.sf.com). Their extensions include advanced functionality out of the box (such as layer rules), as well as customer specific add-ons and a new Graphical User Interface for displaying ModelCHECK results in a more easily read format.

# ABOUT THE AUTHOR

Rosemary L. Astheimer is a Continuing Lecturer for the Polytechnic Institute at Purdue University, in West Lafayette, Indiana. She received her BS in Mechanical Engineering from the University of Massachusetts in Amherst and her Master of Software Engineering from Brandeis University. Before joining the faculty in 2014, Ms. Astheimer spent over 15 years working in the CAD software industry. She began her career in software support, transitioned into a pre-sales applications engineer focusing on business development of new products and was most recently a product manager.

She has held industry positions with 2 of "the big 3" CAD companies, PTC and Siemens PLM Software (formerly VISTAGY), and has had the opportunity to work with aerospace and automotive manufacturing customers. This in-depth experience with major CAD systems, including CATIA V4, V5, NX (formerly Unigraphics), and Creo filters into her instruction style at Purdue, where she teaches CAD design, MBD, PDM and PLM to both undergraduate and industry professionals. In 2016, Ms. Astheimer earned her Technologist level certification as a Geometric Dimensioning & Tolerancing Professional from the American Society of Mechanical Engineers.

In her free time, she enjoys watching Formula 1 racing and has occasionally been seen running her street car around the track at High Performance Driving events with the Sports Car Drivers Association at the New Hampshire International Speedway. She also enjoys sewing (which is basically a specialized type of an assembly made from flexible material!) and spending time with her son. She is an active member and league coordinator of the Executive Women's Golf Association Boston chapter.

# INDEX

# END NOTES

---

[i] http:// dx.doi.org/10.6028/NIST.GCR.15-1009

[ii] http://rand.com/

[iii] https://en.wikipedia.org/wiki/Computervision

[iv] http://www.ptc.com/cad/pro-engineer

[v] Parametric Technology Corporation, http://www.ptc.com

[vi] http://www.rp-ml.org/rp-ml-1998/2502.html

[vii] http://bloomberg.com/research/stocks/private/snapshot.asp?privcapID=31907

[viii] http://sqlmag.com/sql-server/parametric-buys-modelcheck

[ix] https://www.bostonglobe.com/business/2013/01/24/parametric-technology-corp-officially-changes-its-name-ptc/wwNvl3SgU8RNAdo4zxFu3H/story.html

[x] Kenn Hartman, DSA Product Lifecycle Management Consulting

[xi] http://discoverykids.com/articles/why-does-it-rain/

[xii] Image from www.BigActivities.com, colored by Ayrton Astheimer

[xiii] Steven D. Levitt & Stephen J. Dubner, Think Like a Freak (2014)

[xiv] W. Clement Stone, The Success System That Never Fails (1962)

[xv] Kenn Hartman, DSA Product Lifecycle Management Consulting

[xvi] http://answers.microsoft.com/en-us/windows/forum/windows_7-windows_programs/in-windows-7-starter-how-do-i-set-environment/a9847c97-2c34-4aa7-99e9-93cd3e31d188

[xvii] https://www.itechtics.com/customize-windows-environment-variables

[xviii] http://www.phschool.com/eteach/social_studies/2003_05/essay.html

[xix] 

Made in the USA
Middletown, DE
21 September 2016